Identification of Multivariable Industrial Processes
for Simulation, Diagnosis and Control

Yucai Zhu and Ton Backx

Identification of Multivariable Industrial Processes

for Simulation, Diagnosis and Control

With 43 Figures

Springer-Verlag
London Berlin Heidelberg New York
Paris Tokyo Hong Kong
Barcelona Budapest

Yucai Zhu, PhD
IPCOS b.v.
P.O. Box 258
5680 AG Best
The Netherlands

Ton Backx, PhD
IPCOS b.v.
P.O. Box 258
5680 AG Best
The Netherlands

ISBN-13: 978-1-4471-2060-5   e-ISBN-13: 978-1-4471-2058-2
DOI: 10.1007/978-1-4471-2058-2

British Library Cataloguing in Publication Data
A catalogue record for this book is available from the British Library

Library of Congress Cataloging-in-Publication Data
A catalog record for this book is available from the Library of Congress

Typesetting: Camera ready by authors
Printed by Antony Rowe Ltd, Chippenham, Wiltshire
69/3830—543210 Printed on acid-free paper

SERIES EDITORS' FOREWORD

The series *Advances in Industrial Control* aims to report and encourage technology transfer in control engineering. The rapid development of control technology impacts all areas of the control discipline. New theory, new controllers, actuators, sensors, new industrial processes, computing methods, new applications, new philosophies, . . ., new challenges. Much of this development work resides in industrial reports, feasibility study papers and the reports of advanced collaborative projects. The series offers an opportunity for researchers to present an extended exposition of such new work in all aspects of industrial control for wider and rapid dissemination.

The identification of industrial processes is an art based on a set of scientific methods. In those cases where it is impossible or too costly to derive a system model from first principles, identification is an indispensable tool. With the advent of the self-tuning regulator, the identification process actually became part of the controller. Consequently, the system identifier or parameter estimator is now a standard part of the control engineer's toolkit. Professor Backx and Dr. Zhu have drawn from their significant industrial experience to demonstrate the importance of system identification in industrial control. The results from two industrial processes have been used to show the valuable potential, the limitations and the pitfalls of the identification art. We are pleased to welcome this timely and valuable addition to the Series.

April, 1993   M. J. Grimble and M. A. Johnson,
Industrial Control Centre,
Glasgow G1 1QE, Scotland, U.K.

PREFACE

Seeing once is better than hearing hundred times
— *Zhao, Chongguo, Han Dynasty, China*

Testing once is better than seeing hundred times

Systems and control the ry has experienced a continuous development in the last few decades. he state space approach and Kalman filter are the products of the 1960s which made it possible for the first time to solve general linear multivariable control problems. Since 1970 adaptive control theory and techniques have been developed. In the 1980s robust control and H_∞-control of multivariable systems have been developed. Fault detection and diagnosis techniques have also been developed in this period. These new techniques hold high promise for industrial applications and attracted much interest among the academic researchers. The impact of these developments on process industries, however, has been very limited, even negligible. When we visit plants in process industries, we find that a typical modern computer control system is a combination of the computer technology of the 1980s and 1990s and classical PID (proportional, integral and differential) control algorithms which are the restrictive single variable control techniques of the 1940s and 1950s. What an imbalance!

Many possible reasons for this failure of technology transfer can be identified. One important reason is the lack of accurate dynamic models of industrial processes, since all the above mentioned modern techniques are model based and hence need reasonably accurate process models. Another reason is the lack of good communication between the modern control community and the personnel of the process industries.

Process identification is the field of mathematical modelling from experimental data. In this book we first introduce the well known identification methods and then we present a unified approach to multivariable industrial process identification. The intended uses of the identified models are: control system design, simulation (prediction) and fault diagnosis. We concentrate on a large class of industrial processes, with clear model applications in mind.

There is an urgent need for such a volume. Though risky, we dare to predict that, in the 1990s, applications of modern systems and contol theory will extend from the aerospace industries to conventional industries, especially the process industries. The reasons are: worldwide competition, shortage of natural resources and environmental pollution. Under such circumstances industrial companies are forced to increase the quality and flexibility of their production and, at the same time, to decrease the use of energy and materials and to diminish pollution. Cheap and reliable computers have made advanced technologies afford-able.

As mentioned before, applications of modern systems and control theory need accurate process models. For industrial processes, analytical modelling by which one derives models based on first principles, is often not feasible: it is either too costly (sometimes even impossible) or it generates models that are not suitable for the uses of control and simulation. Identification, by which one derives process models from observations and measurements, is certainly a powerful tool for industrial applications of modern systems and control theory.

Existing books on process (system) identification, however, do not meet the needs of industrial control engineers. The authors of such books are mainly researchers from universities, their intended readers are academic researchers and students. The contents of these books are considered "too theoretical" for industrial control engineers. Not enough attention has been paid to model applications, and many practically oriented questions are not answered. Also the contents are organized in a manner which is unhelpful for engineers who have applications in mind: often the experimental design and practical aspects are placed at the end of the book. This forces engineers to study the whole book (which is very time consuming for them) before they can try anything on their processes. A frequent reaction from engineers is to ignore this important topic. This partly explains why process identifi-cation is not widely applied in the process industries, although researchers believe that identification is already a mature science and that it is an effective approach to industrial process modelling.

The purpose of our book is, therefore, to fill the gap between theory and application. We present our identification methods, which can be applied (and has been applid) to multivariable (multi-input multi-output, or MIMO) industrial processes for the purposes of robust controller design, simulation and diagnosis. This work summarizes the results of many years of research and application experiences of our team. Our method is both straightforward and powerful. An engineer who under-stands transfer functions and the least-squares principle, can understand and use the methods. The methods are powerful and they can identify multivariable processes with high accuracy (when compared to existing methods); fundamental problems such as optimal input design and model structure selection can be solved in a straightforward manner and one obtains not only a nominal parametric model, but also error bounds on the transfer function estimates. We have a recursive version of the algorithm and we also propose a simple way for estimating continuous-time models. These nice properties of our methods might possibly attract interest from researchers.

The book is oganized in a way that is reader friendly and easy to use for engineers. Instead of the usual top-down approach, i.e., first build a very general framework and then showing that different methods are special cases of the general theory, we will start with the simplest method, and then gradually introduce other methods, i.e., the bottom up approach. In this way, we can bring more physical insight to our readers and much mathematics can be avoided. A reader need not jump between different chapters to use a specific identification method, because each method is treated in a single chapter or section and experiment design is explained before discussing any identification algorithms. Model fit in the frequency domain is a central issue of the book, which is reflected in studying the existing methods and in developing our new methods. This way of thinking is of course very natural for control engineers. Last but not least, actual industrial tests will show the power of process identification and make the theory less abstract.

Preliminary knowledge needed for reading our book are an introduction to linear systems theory and digital control (see, e.g., Åström and Wittenmark, 1984), and some knowledge about matrix operations, probability and random variables (see, e.g., Eykhoff, 1974, Appendixes B, C, and D).

Yucai Zhu, PhD
IPCOS b.v.
P.O. Box 258
5680 AG Best
The Netherlands

Ton Backx, PhD
IPCOS b.v.
P.O. Box 258
5680 AG Best
The Netherlands

Acknowledgements

The main part of our results in the book were obtained during our PhD research at the Measurement and Control Group, Department of Electrical Engineering, Eindhoven University of Technology, The Netherlands, under the supervision of Prof. Pieter Eykhoff. We thank him for his guidance and support, especially for his encouragement to pursue applications orientated research in process identification. Besides, he was so kind as to provide a review of this book. We also wish to thank Dr. Ad Damen, with whom we had, and still have, many technical discussions. His clear mind and sharp views have stimulated many new ideas. We also thank our former colleagues Dr. Ad van den Boom, Dr. Andrej Hajdasinski, Dr. Paul van den Hof, Dr. Peter Janssen, for their help and inspiring discussions.

Much of the practical results and experience come from the industrial projects carried out by our colleagues at IPCOS b.v., and we are grateful for their contributions to the book. IPCOS b.v. was formerly the IPCOS Unit of Datex b.v., and we thank Datex b.v. for their financial support of the IPCOS activities.

Ton Backx was with Philips Glass (which now belongs to Philips Lighting) when he initiated a research project on process identification. He wants to thank the direction and colleagues at Philips Glass. In particular he would like to thank Mr. Bert van den Braak for his stimulation and leadership.

The study of Yucai Zhu in Eindhoven was made possible by a scholarship of Xi'an Jiaotong University, China. He is very grateful for this opportunity. He also wishes to express his gratitude to Mr. Tan S. Liong and his family. With family Tan, he enjoys family life while he is 10,000 km away from his parents.

We would like to thank the editors of the series Prof. Mike Grimble and Dr. Mike Johnson for their encouragement and support. The careful review of Dr. Johnson has improved the quality of this volume both in language and in technical contents. However, the authors remain fully responsible for the mistakes that remain in the text.

Finally we want to thank our families for their support, patience and the joy of life.

CONTENTS

CHAPTER 1

INTRODUCTION

1.1   Some Preliminary Concepts

System.   A system is a collection of objects arranged in an ordered form to serve some purpose. Everything not belonging to the *system* is part of the *environment*. One may characterize the system by input/output (cause and effect) relations. What constitutes a *system* depends on the point of view of the observer. The *system* may be, for example, an amplifier consisting of electronic components, or a control loop including that amplifier as one of its parts, or a chemical processing unit having many such loops, or a plant consisting of a number of units or a number of plants operating together as a system in the environment of a global economy.

Process.   A process is a *processing system* that serves some purposes in which some variables or quantities are experiencing continuous change and evolution. In a process different kinds of variables interact and produce observable variables. The observable variables of interest to us are usually called *outputs*. The process is also affected by external variables. External variables that can be manipulated by us are *inputs* of the process. Other external variables are called *disturbances*. Disturbances can be divided into two kinds: the measurable disturbances which can be directly measured, and the unmeasurable disturbance which are only observed through their influence on the outputs. A process is said *dynamic* when the current output value depends not only on the current external stimuli but also on their earlier values. Outputs of dynamic processes whose external variables are not observed are often called *time series* (Ljung, 1987).

Model.   A model is a representation of the essential aspects of a system (process) which presents knowledge of that system (process) in a usable form (Eykhoff, 1974).

There are many types of models. People are most familiar with *mental models* which do not involve any mathematical formalization. To drive a car, for example, the driver has a mental model about the relations between the turning of the car direction and the turning of the steering wheel, and between the acceleration of the car and the accelerator. For manual control of a industrial process, the process operator needs the knowledge about how the process outputs will respond to various control actions. Sometimes it is appropriate to describe the properties of a system by means of tables or plots. Such descriptions are called graphical models. Bode plots, step responses and impulse responses of linear systems are of this type.

For the application of modern systems and control theory, it is necessary to use *mathematical models* that describe the relationships among the system variables in terms of difference or differential equations. In fact, the use of mathematical models is not limited to the control community; a major part of the engineering field deals with how to use mathematical models for simulations, forecasting and designs. The topic of this book— system identification—is about how to obtain mathematical models of systems (processes) from observations and measurements.

System Identification. System or process identification is the field of mathematical modelling of systems (processes) from experimental data. In technical terms, *system identification* is defined by Zadeh (1962) as: *the determination on the basis of input and output, of a system (model) within a specified class systems (models), to which the system under test is equivalent (in terms of a criterion).*

It follows from this definition that three basic entities are involved in system identification: *the data,* a *model set* and a *rule* or *criterion* for model estimation.

The input/output data are usually collected from an identification experiment that is designed to make the measured data maximally informative about the system properties that are of interest to the user.

A set of candidate models is obtained by specifying their common properties, a suitable model is searched for within this set. This is the most theoretical part of the system identification procedure. It is here that *a priori* knowledge and engineering intuition and insight have to be combined with the formal (mathematical) properties of models. In this book we will restrict ourselves to linear, time-invariant and finite dimensional models of multi-input multi-output (MIMO) systems that are suitable for modelling a large class of industrial processes.

When the data are available and the model set is determined, the next step is to find the best model from the data. To measure the goodness of the model fit a criterion (loss function) needs to be specified. Often the sum of the squares of some error signals (residuals) is used as the criterion.

The system identification procedure has the following basic steps:

1) Identification experiments.

2) Model order/structure selection.

3) Parameter estimation.

4) Model validation.

Model validation is the process of testing whether the estimated model is sufficiently good for the intended use of the model. First of all, a check to see if the model is in agreement with the *a priori* knowledge of the system. Then a check if the model can fit the experimental data well, preferably using a data sequence that has not been used in model estimation. The final validation of the model is the application of the model. Fig. 1.1.1 shows the identification procedure; see Ljung (1987).

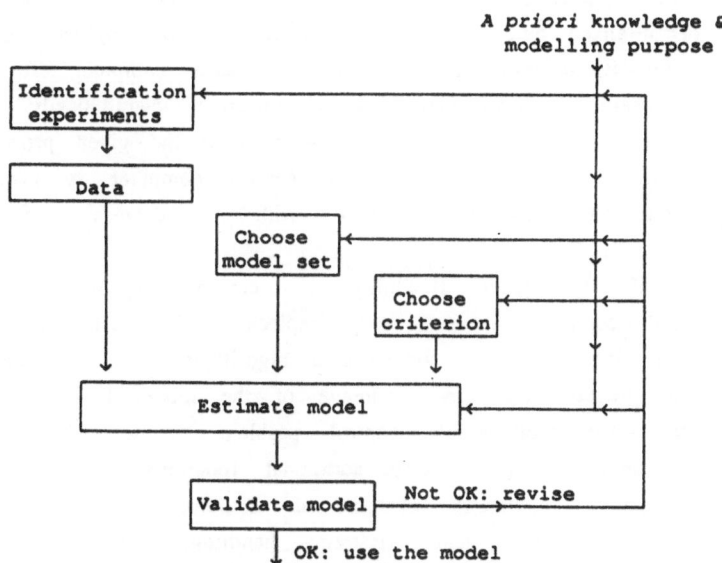

Fig. 1.1.1 The identification procedure

It is possible that a model first obtained cannot pass the model validation procedure. The causes of the model deficiency can be the following:

— The data were not informative enough due to poor experiment design.

— The model set did not contain a good description of the system.

— The criterion was not chosen properly.

— The numerical procedure failed to find the best model according to the criterion.

The major part of an identification application consists of addressing these problems. A good identification method, therefore, should provide systematic solutions to these problems.

1.2 Digital Control of Industrial Processes

Currently, powerful and cost effective digital processors and data storage devices determines the present development of process control. Nowadays, by means of microprocessors which exceed the capacity of former process computers, decentralized automatic control systems can be applied. To do so, the tasks previously centrally processed in a process computer are delegated to various process microcomputers. Many different hierarchically organized structures can be build up. They can be adapted to the given process. In a decentralized system the high load of a central computer is avoided and operational reliability increases together with a decrease of software complexity.

Because of the great flexibility of control algorithms stored in software, digital controllers cannot only replace analog controllers of PID type, they can also use more sophisticated algorithms based on mathematical models of the process. The key advantage of the model based controller is its ability to solve multivariable control problem as we shall see later. Besides pure control function, many additional functions can be introduced, such as adaptation, automatic change of setpoint, fault detection and diagnosis, online simulation and constraint handling. These new functions are urgently needed by process industries in order to improve product quality, to increase production flexibility, to reduce the cost of the material and energy usage and minimise environmental impact. At present this is still a dream for most process control engineers, for we rarely find a digital controller that is equipped with all these functions.

We hope that the application of modern systems and control theory based on precise process models will make the dream become true in the very near future.

Examples of industrial processes that have great potential to benefit from the application of modern systems and control theory are, among others:

— Industrial furnaces (glass, steel, etc.);

— distillation columns;

— polymerization reactors;

— rolling mills;

— glass forming processes;

— drying processes (spray dryers, evaporators);

— biotechnology processes.

1.3 Outline of the Book

In this book we first introduce the existing methods of linear process identification, then present our own methods that we have developed over the last decade.

In Chapter 2 we introduce several different mathematical representations of linear dynamic processes and signals which are necessary in process identification and control system design.

After laying down the basic concepts for system models, we start immediately, in Chapter 3, with the identification experiments. This is the first and the most crucial step in a identification procedure. In this chapter, problems such as the selection of inputs and outputs, sampling frequency, test signals and pre-treatment of the data are discussed in detail. Different experiments and tests are suggested for gathering the necessary information about the process dynamics. At this stage, we intend our readers to be able to design *good* experiments, rather than *optimal* ones. The material presented here is based on experience, classical control engineering and common sense.

When input/output data are collected, we suggest our reader begins with the simplest method—*least-squares* method (also called *equation error* method or ARX model method). This method is treated in Chapter 4. The least-squares method is most commonly used in applications, and many control engineers have some knowledge about it. After the introduction of the method, we present two industrial cases in order to illustrate the method. In the first case, the identification and control of a single stand rolling mill, very successful results are obtained. This will give our reader the first taste of the usefulness of identification. He can try the same thing on his own process. If he succeeds, he can stop reading the book and continue his other (more important) business. If he has the bad luck that least-squares method does not work, just like we have had in the second case, the subsequent sections will explain how and why the least-squares method fails. Also, if a reader wants to obtain better results than those of the least-squares method, he will be motivated to read the following sections and chapters.

In Chapter 5 different ways to overcome the bias problem of the least-squares method are explained. Several popular techniques, such as the output error method, instrumental variable methods and prediction error methods are introduced. Their properties are explained in parametric and in frequency response terms. Also practical limitations of this class of methods will be pointed out. The shortcomings of these methods call for other alternatives.

In Chapter 6 we present the Markov parameter (finite impulse response) approach to MIMO process identification. The method uses the output error criterion. In order to find a good initial estimate and to determine the order of the model, first a finite impulse response (FIR) model or Markov parameters are estimated by linear least-squares method. Then the order of the model is determined by examining the singular values of a special Hankel matrix of the Markov parameters. A model reduction technique is used to fit a reduced model to the FIR model. Finally the low order model is adjusted to the original data. Thus the method is numerically robust; and difficult problem of MIMO model structure selection is avoided here. The method is applied on a glass tube production process.

In Chapter 7, we present the *two-step* approach. The method is based on the frequency domain measures. It consists of a least-squares estimation of high order model and subsequent model reduction. In doing so, we can easily solve fundamental problems such as optimal input design and model order determination; the numerical algorithms involved are mainly least-squares and modified least-squares. One important feature of our method is its ability in supplying an upper bound of model errors in the frequency domain, which is needed for robust control design. Also recursive estimation is discussed here and we show that our method can be rewritten in a recursive form. Simulation studies will be used to validate the method and to make comparisons with other methods.

In Chapter 8 we extend the two-step method for use with multivariable processes. We will not bother ourselves with minimal canonical forms of multivariable process descriptions which is difficult to learn and imprac-tical to use. A state space realization of the overall model is obtained by performing balanced model reduction on the diagonal model form. Finally, this method delivers a parametric model in a form of polynomial matrix or a state space realization, together with an upper bound matrix for the model errors. Two industrial processes, the glass tube process and a multi-effect evaporator, will be identified using the method; the results will be compared with those obtained using some prediction error methods. Finally we show how to use the method in the closed loop identification of (possibly unstable) MIMO processes.

Chapter 9 describes an application of identification and control of the glass tube manufacturing process. For solving the problem our identification method is combined with some recent control theory. In this chapter, the reader can learn more on how to integrate identification and control, and the successful application may encourage him to use identification to solve his own problems.

In Chapter 10 we study the problem of continuous-time identification which is needed for model based fault detection and diagnosis. First we propose an indirect approach to continuous-time model estimation. Then a method of input design for enhancing the accuracy of part of the parameters is proposed. The methods are tested in a simulation study.

CHAPTER 2

LINEAR MODELS OF DYNAMIC PROCESSES AND SIGNALS

Process identification is the subject of constructing models for certain utilitarian purposes from measured data. A first step in the total identification procedure is to determine a class of models in which the most suitable model is to be sought. In this chapter we will review some well known results on how to describe dynamic processes and signals using linear models.

In this book, we only discuss the identification of linear time-invariant models of industrial processes. Linear models form the most important and well developed class of models both in practice and in theory. Although they represent idealized processes in the real world, the approximations involved are often justified for the given applications of the models. A typical example is to describe the behaviour of a industrial process, possibly nonlinear, by a linear time-invariant model around a working point. We first discuss continuous-time models in Section 2.1; then discuss discrete-time models in Section 2.2. Models of multi-input multi-output processes are discussed in Section 2.3. Models of signals are discussed in Section 2.4. Section 2.5 is the concluding section of the chapter.

2.1 SISO Continuous-Time Models

Most of the industrial processes we study are inherently continuous-time processes; all the signals of this kind of processes are time-continuous. So it is logical to start with continuous-time models. In this book, we use the models that are described by linear ordinary differential equations connected with pure time delays.

A linear time-invariant single-input single-output process can be described by an ordinary differential equation as

$$y^{(n)}(t) + a_1 y^{(n-1)}(t) + \cdots a_n y(t) = b_0 u^{(m)}(t) + b_1 u^{(m-1)}(t) + \cdots b_m u(t)$$

$$(2.1.1)$$

where $u(t)$ and $y(t)$ are the input and the output at time t; superscript (n)

denotes n-th order derivative. The coefficients a_1, $\cdots$ a_n, b_0, $\cdots$ b_m are constant. The order of the process model is defined as n. Note that $n \geq m$ for a physically realizable or causal process.

Taking the Laplace transform on both sides of (2.1.1) and assuming zero initial conditions yields the transfer function of the process model:

$$G(s) = \frac{Y(s)}{U(s)} = \frac{b_0 s^m + b_1 s^{m-1} + \cdots + b_m}{s^n + a_1 s^{n-1} + \cdots + a_n} \tag{2.1.2}$$

where $U(s)$ and $Y(s)$ are the Laplace transforms of the input and the output.

A linear time-invariant process with measurement delay d is described by a differential equation as

$$y^{(n)}(t) + a_1 y^{(n-1)}(t) + \cdots a_n y(t)$$
$$= b_0 u^{(m)}(t-d) + b_1 u^{(m-1)}(t-d) + \cdots b_m u(t-d) \tag{2.1.3}$$

and by the transfer function as

$$G(s) = \frac{Y(s)}{U(s)} = \frac{b_0 s^m + b_1 s^{m-1} + \cdots + b_m}{s^n + a_1 s^{n-1} + \cdots + a_n} e^{-sd} \tag{2.1.4}$$

It is well known that a linear time-invariant and causal process can be described by impulse response $g(\tau)$ as follows

$$y(t) = \int_0^\infty g(\tau) u(t-\tau) d\tau \tag{2.1.5}$$

If we introduce a vector of auxiliary variables and express (2.1.1) as a set of first order differential equations, we get a state space realization of the model:

$$\dot{x}(t) = Ax(t) + Bu(t)$$
$$y(t) = Cx(t) + Du(t) \tag{2.1.6}$$

where $x(t) = [x_1(t) \cdots x_n(t)]^T$ is called a state space vector. Assume that $m = n-1$ in (2.1.1), then a state space realization of the process is

$$A = \begin{bmatrix} -a_1 & 1 & 0 & \cdots & 0 \\ -a_2 & & 1 & & \\ \vdots & & & \ddots & \\ & & & & 1 \\ -a_n & 0 & 0 & \cdots & 0 \end{bmatrix}, \quad B = \begin{bmatrix} b_0 \\ b_1 \\ \vdots \\ b_{n-1} \end{bmatrix}$$

(2.1.7)

$$C = [1 \ 0 \ \cdots \ 0], \quad D = [0 \ 0 \ \cdots \ 0]$$

This is called the observer canonical form in the literature; see, e.g., Kailath (1980).

A given differential equation can have many state space descriptions. Therefore state space realizations are not unique. A great advantage of using state space models is their convenience for analysis and controller design for MIMO processes.

It is well known that the transformation from a state space description to the transfer function is

$$G(s) = B(sI - A)^{-1}C + D$$

(2.1.8)

2.2 SISO Discrete-Time Models

A schematic diagram of a computer control system is given in Fig. 2.2.1. Due to the use of process computers, signals from the underlying continuous-time process are sampled and digitized. These signals have discrete amplitude values at discrete-time points. Models describing the relationship between these kind of signals are called discrete-time models.

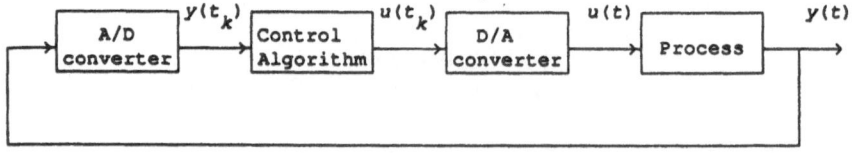

Fig. 2.2.1 Schematic diagram of a computer control system

In this book, we assume synchronous sampling at computer inputs and outputs and under a single sampling rate (frequency); also, we assume that the quantization error is much smaller than the measurement noises and disturbances such that the sampled amplitudes can be considered continuous. Under these realistic assumptions, a sampled linear process can be described by a difference equation or a z-transfer function relationship.

Typically a discrete-time description of a sampled continuous-time process is obtained as follows. In most of the computer controlled systems, the process input $u(t)$ (the output of the D/A converter) is kept constant during a sampling interval using a zero order hold. This practical arrangement also greatly simplifies the theoretical analysis. Suppose the process is linear and time-invariant with impulse response $g(\tau)$, then at sampling instants $t_k = kT$, $k = 1, 2, \cdots$, where T is the sampling interval:

$$y(kT) = \int_0^\infty g(\tau)u(kT-\tau)d\tau \tag{2.2.1}$$

Because the input is constant over the sampling interval (piecewise constant), then

$$u(t) = u_k, \qquad kT \le t < (k + 1)T \tag{2.2.2}$$

(2.2.1) becomes

$$y(kT) = \int_0^\infty g(\tau)u(kT-\tau)d\tau = \sum_{l=1}^\infty \int_{\tau=(l-1)T}^{lT} g(\tau)u(kT-\tau)d\tau$$

$$= \sum_{l=1}^\infty \left[\int_{\tau=(l-1)T}^{lT} g(\tau)d\tau \right] u_{k-l} = \sum_{l=1}^\infty g_T(l)u_{k-l} \tag{2.2.3}$$

where we defined

$$g_T(l) = \int_{\tau=(l-1)T}^{lT} g(\tau)d\tau, \qquad l = 1, 2, \cdots \tag{2.2.4}$$

which is called the discrete-time impulse response of the sampled process. Note that no approximation is involved in deriving (2.2.3).

Even if the input is not piecewise constant, the representation (2.2.3) may still be a reasonable approximation provided $u(t)$ does not change too much during a sampling interval.

We will use the notation (2.2.1) to (2.2.4) when the choice of sampling interval is discussed. In most of the Chapters we will study the identification of discrete-time processes. For ease of notation we shall assume that the sampling interval is one time unit and we use t to denote the sampling instants. This yields for (2.2.3)

$$y(t) = \sum_{k=1}^{\infty} g_k u(t-k), \qquad t = 0, 1, 2, \cdots \tag{2.2.5}$$

Let us introduce the *unit forward shift operator* q

$$qu(t) = u(t+1)$$

and the *unit delay operator* q^{-1}

$$q^{-1}u(t) = u(t-1)$$

We can then write for the convolution sum of (2.2.5)

$$y(t) = \sum_{k=1}^{\infty} g_k [q^{-k} u(t)] = \left[\sum_{k=1}^{\infty} g(k) q^{-k} \right] u(t) \tag{2.2.6}$$

Define the *transfer operator* of the process (2.2.5) from the convolution of (2.2.6) as:

$$G(q) = \sum_{k=1}^{\infty} g_k q^{-k} \tag{2.2.7}$$

We then obtain a convenient description of the process as:

$$y(t) = G(q)u(t) \tag{2.2.8}$$

It is known that the sampling of an n-th order linear time-invariant process using a zero-order hold results in an n-th order difference equation as follows

$$y(t) + a_1 y(t-1) + \cdots a_n y(t-n) = b_1 u(t-1) + \cdots b_n u(t-n) \tag{2.2.9}$$

Note that the coefficients here have different meanings to those of (2.1.1), although the symbols are the same. This will not cause confusion, because we only treat discrete-time models in the book except in Chapter 10. Note also that there is one delay in the difference equation which is caused by the zero-order hold.

The transfer operator of (2.2.9) is

$$G(q) = \frac{b_1 q^{-1} + \cdots + b_n q^{-n}}{1 + a_1 q^{-1} + \cdots + a_n q^{-n}} \tag{2.2.10}$$

If there is a delay of d samples, then the transfer operator becomes

$$G(q) = \frac{b_1 q^{-1} + \cdots + b_n q^{-n}}{1 + a_1 q^{-1} + \cdots + a_n q^{-n}} \cdot q^{-d} \tag{2.2.11}$$

We find that in discrete-time, a finite dimensional model with a delay is still finite dimensional, while in continuous-time, a delay cannot be modelled exactly by a finite dimensional model; compare (2.1.4). This is an advantage of using discrete-time models.

Remarks— If we replace q by the z-transform variable z, then $G(z)$ is called transfer function of the process. If we evaluate the transfer function on the unit circle, $z = e^{i\omega}$, then $G(e^{i\omega})$ is called the frequency response of the process. Sometimes people do not observe the difference between q and z, which is theoretically not correct, but this will not do any harm in applications.

— The leading coefficient of the denominator polynomial of the transfer operator (2.2.10) is fixed as 1 for the uniqueness of the model. A polynomial is called monic if its leading coefficient is unity.

— Models in the form of impulse response (2.2.5) are called non-parametric models; models in the form of difference equation (2.2.9) are called parametric models.

If we rearrange (2.2.9) as a set of first order difference equations, we get a state space realization of the model:

$$\begin{aligned} x(t+1) &= Ax(t) + Bu(t) \\ y(t) &= Cx(t) + Du(t) \end{aligned} \tag{2.2.12}$$

where $x(t) = [x_1(t) \cdots x_n(t)]^T$ is a state space vector. The observer canonical form of (2.2.9) is given as

$$A = \begin{bmatrix} -a_1 & 1 & 0 \cdots & 0 \\ -a_2 & & 1 & \\ \vdots & & & 1 \\ -a_n & 0 & 0 \cdots & 0 \end{bmatrix}, \quad B = \begin{bmatrix} b_1 \\ b_2 \\ \vdots \\ b_n \end{bmatrix} \tag{2.2.13}$$

$$C = [1 \ 0 \ \cdots \ 0], \quad D = [0 \ 0 \ \cdots \ 0]$$

The transfer operator can be obtained from a state space description by

$$G(q) = B(qI - A)^{-1}C + D \qquad (2.2.14)$$

2.3 MIMO Models

Because we will mainly treat discrete-time models in this book and there are many similarities between continuous-time models and discrete-time models, we shall only study discrete-time MIMO models in this section.

Given a discrete-time linear time-invariant process with m inputs, p outputs, we can describe the relation between the inputs and the outputs by impulse responses and convolutions

$$y(t) = \sum_{k=1}^{\infty} G_k u(t-k), \qquad t = 0, 1, 2, \cdots \qquad (2.3.1)$$

where $y(t)$ is the p-dimensional output vector and $u(t)$ is the m-dimensional input vector at sampling instant t; G_k is a sequence of $p \times m$ matrices which form the discrete-time impulse response. Using the unit time delay operator q^{-1} we obtain

$$y(t) = G(q)u(t) \qquad (2.3.2)$$

where

$$G(q) = \sum_{k=1}^{\infty} G_k q^{-k} \qquad (2.3.3)$$

is called transfer operator of the MIMO process.

If the process is finite dimensional, then each entry of the transfer operator matrix $G(q)$ can be written as a rational function of q^{-1} in the form of (2.2.10) or (2.2.11). This form is not convenient for identification and also not suitable for controller design.

For the purpose of identification, a suitable description of a MIMO process is a set of difference equations as follows

$$y(t) + A_1 y(t-1) + \cdots + A_n y(t-n) = B_1 u(t-1) + \cdots + B_n u(t-n) \qquad (2.3.4)$$

where A_1 $(p \times p)$, $\cdots A_n$ $(p \times p)$, B_1 $(m \times p)$ $\cdots B_n$ $(m \times p)$ are constant matrices.

Using the unit delay operator q^{-1}, we write for (2.3.4)

$$A(q)y(t) = B(q)u(t) \qquad (2.3.5)$$

where $A(q)$ and $B(q)$ are polynomial matrices:

$$A(q) = I + A_1 q^{-1} + \cdots + A_n q^{-n}, \qquad B(q) = B_1 q^{-1} + \cdots + B_n q^{-n}$$

Comparing (2.3.2) and (2.3.5) we have

$$G(q) = A(q)^{-1}B(q) \qquad (2.3.6)$$

provided that $A(q)$ is invertible. Expressing the transfer operator matrix by the two polynomial matrices is called *matrix fraction description* (MFD). The parameters of matrices $A(q)$ and $B(q)$ can be estimated from the data, because they are directly related to the inputs and outputs. Note that the degrees of the entries of $A(q)$ and $B(q)$ can be lower than n.

Unlike the SISO case, for a given process the representation (2.3.5) is in general not unique even when we fix the leading coefficient matrix of $A(q)$ at I. Therefore some special forms of representations (canonical forms) are needed in order to have unique models of a given process; and model uniqueness is a necessary condition for an identification algorithm to find a meaningful solution. Canonical forms for MIMO processes are often regarded as difficult. This difficulty can be avoided by using one of the simplest canonical form, *diagonal form* MFD. The diagonal form MFD of a given $G(q)$ is

$$A(q) = \begin{bmatrix} A_{11}(q) & 0 & \cdots & 0 \\ 0 & A_{22}(q) & & \vdots \\ \vdots & & \ddots & 0 \\ 0 & \cdots & 0 & A_{pp}(q) \end{bmatrix}, \quad B(q) = \begin{bmatrix} B_{11}(q) & \cdots & B_{1m}(q) \\ \vdots & \ddots & \vdots \\ B_{p1}(q) & \cdots & B_{pm}(q) \end{bmatrix} \qquad (2.3.7)$$

where $A_{11}(q)$, $\cdots$, $A_{pp}(q)$ are all monic polynomials with relevant degrees, and the degrees of $B_{11}(q)$, $\cdots$, $B_{im}(q)$ are equal to or less than that of $A_{ii}(q)$. Under this arrangement, a m-input, p-output process is decoupled into p m-input single output subprocesses

$$A_{11}(q)y_1(t) = B_{11}(q)u_1(t) + \cdots + B_{1m}(q)u_m(t)$$
$$\vdots \qquad \vdots \qquad \vdots \qquad (2.3.8)$$
$$A_{pp}(q)y_p(t) = B_{p1}(q)u_1(t) + \cdots + B_{pm}(q)u_m(t)$$

Any given transfer operator of a physical process can be written in the diagonal form MFD. This form has many advantages for use in identification:

— It is simple to comprehend;

— using this form, (almost) all the identification algorithms developed for SISO processes (see Chapters 5 and 6) can be generalized for MIMO processes;

— delay correction (see Chapter 3) can be done for each single transfer operator.

The only possible disadvantage is that the diagonal form MFD of a given process is not always minimal. This means that the order of its direct state space realization can be higher than the McMillan degree of the process. This problem, however, can be solved using some model reduction techniques; see Chapter 8.

In recent years, the Hankel matrix has played a very important role in approximate modelling (identification, model reduction) and robust control system design (H_∞ control). The Hankel matrix $\mathcal{H}$ of a discrete-time process $G(q)$ is defined as the double infinite matrix

$$\mathcal{H} := \begin{bmatrix} G_1 & G_2 & G_3 & \cdots \\ G_2 & G_3 & G_4 & \cdots \\ G_3 & G_4 & G_5 & \cdots \\ \vdots & \vdots & \vdots & \end{bmatrix} \tag{2.3.9}$$

where $\{G_k\}_{k=1,\cdots,\infty}$ is the impulse response of $G(e^{i\omega})$.

Denote δ as the minimal order (Mcmillan degree) of $G(q)$, then it is well known that

$$rank(\mathcal{H}) = \delta \tag{2.3.10}$$

The δ singular values of $\mathcal{H}$, $h_1(G)$, $\cdots$, $h_\delta(G)$, are called Hankel singular values (HSV) of the process $G(q)$; and, by convention, they are ordered as

$$h_i(G) \geq h_{i+1}(G) \qquad \forall i \tag{2.3.11}$$

where $h_1(G)$ is also called the *Hankel norm* of $G(q)$.

2.4 Models of Signals

Time domain variables such as inputs, states, outputs and disturbances of a process, are called signals. In this section we will discuss how to describe signals using a frequency domain representation and also a probability framework. Only SISO discrete-time case is studied.

Periodograms of Signals over Finite Intervals

Consider the finite sequence of input $u(t)$, $t = 1, 2, \cdots, N$. Define the function $U_N(\omega)$ by

$$U_N(\omega) = \frac{1}{\sqrt{N}} \sum_{t=1}^{N} u(t) e^{-i\omega t} \qquad (2.4.1)$$

The values at $\omega = 2\pi k/N$, $k = 1, \cdots, N$, form the discrete Fourier transform (DFT) of the sequence. The time domain signal can be obtained by the inverse DFT as

$$u(t) = \frac{1}{\sqrt{N}} \sum_{k=1}^{N} U_N(\frac{2\pi k}{N}) e^{-i2\pi kt/N} \qquad (2.4.2)$$

We know that $U_N(\omega)$ is periodic with period 2π.

It is well known by control engineers that DFT decomposes the time domain signal into its frequency domain components; the number $U_N(2\pi k/N)$ is the weight at $\omega = 2\pi k/N$. The value $|U_N(\omega)|^2$ is called *periodogram* of the signal $u(t)$, $t = 1, \cdots, N$; and $|U_N(2\pi k/N)|^2$ is a measure of the energy contribution of the signal at frequency $\omega = 2\pi k/N$.

The energy of the signal $u(t)$, $t = 1, \cdots, N$ can be determined in both the time domain and the frequency domain, using the well known and important Parseval's formula

$$\sum_{k=1}^{N} |U_N(\frac{2\pi k}{N})|^2 = \sum_{t=1}^{N} |u(t)|^2 \qquad (2.4.3)$$

Signal Spectra

Let $\{v(t)\}$ be a stationary stochastic process (see Eykhoff, 1974, Appendix B), the autocorrelation function of it is defined by

$$R_v(\tau) = Ev(t)v(t-\tau) \tag{2.4.4}$$

where E denotes mathematical expectation, and the (power) spectrum (or spectral density) of it is defined as

$$\Phi_v(\omega) = \sum_{\tau = -\infty}^{\infty} R_v(\tau)e^{-i\tau\omega} \tag{2.4.5}$$

Denote $\{w(t)\}$ as another stationary stochastic process, the crosscorrelation function between $\{v(t)\}$ and $\{w(t)\}$ is defined as

$$R_{vw}(\tau) = Ev(t)w(t-\tau) \tag{2.4.6}$$

and the cross spectrum between the two is

$$\Phi_{vw}(\omega) = \sum_{\tau = -\infty}^{\infty} R_{vw}(\tau)e^{-i\tau\omega} \tag{2.4.7}$$

By definition of the inverse Fourier transform, we have

$$Ev^2(t) = \frac{1}{2\pi}\int_{-\pi}^{\pi} \Phi_v(t)d\omega \tag{2.4.8}$$

Let $G(q)$ be a stable transfer operator (filter) and let

$$s(t) = G(q)v(t) \tag{2.4.9}$$

Then $\{s(t)\}$ is also a stationary process and

$$\Phi_s(\omega) = |G(e^{i\omega})|^2\Phi_v(\omega) \tag{2.4.10}$$

$$\Phi_{sv}(\omega) = G(e^{i\omega})\Phi_v(\omega) \tag{2.4.11}$$

For a given sequence (a realization), $v(t)$, $t = 1, \cdots, N$, of the process $\{v(t)\}$, the autocorrelation function can be estimated by replacing the expectation by the time averaging operator:

$$\hat{R}_v(\tau) = \frac{1}{N}\sum_{t=1}^{N} v(t)v(t-\tau) \tag{2.4.12}$$

Then the spectrum can be calculated as

$$\hat{\Phi}_v(\omega) = \sum_{\tau = -\tau_m}^{\tau_m} \hat{R}_v(\tau)e^{-i\tau\omega} \tag{2.4.13}$$

with a suitable τ_m, for example, $\tau_m = N/10$. It can be shown that when $N \to \infty$

these two estimates will converge with probability one to $R_v(\tau)$ and $\Phi_v(\omega)$, provided that the process $\{v(t)\}$ is *ergodic* (see Eykhoff, 1974, Appendix B). This implies that when the number of samples N is large, these estimates will be accurate. The crosscorrelation function and the cross spectrum can be estimated similarly.

For a deterministic signal $u(t)$ we can define the correlation function and spectrum by

$$R_u(\tau) = \lim_{N\to\infty} \frac{1}{N} \sum_{t=1}^{N} u(t)u(t-\tau) \qquad (2.4.14)$$

$$\Phi_u(\omega) = \sum_{\tau=-\infty}^{\infty} R_u(\tau)e^{-i\tau\omega} \qquad (2.4.15)$$

provided that they exist.

If the signal $\{u(t)\}$ is partly deterministic and partly random, Ljung (1987) combined the definitions (2.4.4) and (2.4.14) to give the autocorrelation function as

$$R_u(\tau) = \lim_{N\to\infty} \frac{1}{N} \sum_{t=1}^{N} Eu(t)u(t-\tau) \qquad (2.4.16)$$

If this function exists, $\{u(t)\}$ is called a *quasi-stationary*. It can be shown (Ljung, 1987) that the fundamental relations (2.4.10) and (2.4.11) are also valid for quasi-stationary processes. The introduction of the principle of quasi-stationary is important for the analysis in process identification. In practice, however, (2.4.12) and (2.4.13) can always be used to estimate $R_u(\tau)$ and $\Phi_u(\omega)$ without bothering about whether it is deterministic or random.

Let the stationary stochastic process $\{v(t)\}$ be generated by filtering a white noise sequence

$$v(t) = H(q)e(t) \qquad (4.2.17)$$

where $\{e(t)\}$ is a sequence of independent random variables with a certain (unknown) probability density function; it has zero mean and variance R; $H(q)$ is a stable and minimum-phase filter (its inverse is stable) which is monic $(H(0) = 1)$. For practical use, we parameterize $H(q)$ as a rational function of q. This description is often a reasonable characterization of random disturbances for practical purposes. According to (2.4.10) we have

$$\Phi_v(\omega) = |H(e^{i\omega})|^2 R \qquad (2.4.18)$$

2.5 Linear Processes with Disturbances; Conclusion

We have discussed how to describe linear processes (or linear models of processes) in the previous sections in a extremely ideal situation, namely the noise-free case. In the modelling of industrial processes, there are always variables or signals beyond our control whose effects on the process cannot be neglected; typically they are measurement noises and unmeasurable disturbance; see the examples in Chapters 4 and 8. For a linear process, the effects of these variables can be lumped into an additive term (or vector) $v(t)$ at the output(s):

$$y(t) = G(q)u(t) + v(t) \qquad (2.5.1)$$

The disturbance vector $v(t)$ is typically not measurable, and is only noticeable via its effect on the outputs. Classical ways of describing disturbances are in the forms of steps and pulses. In identification and stochastic control, the disturbances are modelled as realizations of stochastic processes as in the previous section. The input vector $u(t)$ can be considered as noise-free, because it is often the test signal and the control actions generated by the computer. Sometimes the process inputs are measured, but the effect of the measurement noise on the outputs is often much smaller than that of the disturbances.

In the coming chapters we will study how to obtain a model of an industrial processes which has the form (2.5.1) and is derived from experimental data.

CHAPTER 3

IDENTIFICATION EXPERIMENTS AND DATA PRE-TREATMENT

Identification is modelling based on experiments. If the experiments are not performed properly then no identification algorithm however sophisticated can arrive at a relevant model of the process. On the other hand, if the process input/ output data are collected from a well designed experiment, the simplest least-squares method can often deliver a good model. Therefore, experiments are the most important part in the total identification procedure. In the identification literature, much more attention has been paid to estimation algorithms than to the design of identification experiments. This often leads people to think that identification is just a set of algorithms and it is simply *data in-and-model out*.

The purpose of experiments is to collect relevant information about the process dynamics and its environment (disturbances). This information is then transformed to mathematical models of the process and of the disturbances by some identification algorithms (see Chapters 4, 5, 6, 7 and 8). Usually several different types of experiments need to be performed; each of them is for collecting certain kind of information about the process which is needed for the next experiment. The model is estimated from the final experiment where the process inputs are continuously perturbed by some test signals.

In this chapter we will describe a sequence of experiments, where we start with very little *a priori* knowledge of the process and end up with the input/output data which can be used for deriving relevant models of the process and of the disturbances. Some related problems such as sampling frequency, anti-aliasing filter, persistent excitation, and pre-treatment of data, will be discussed. In Section 3.1 we will motivate the different steps of the experiments; the choice of process inputs and outputs are discussed; and several preliminary experiments for collecting *a priori* information of the process are outlined; in Section 3.2 we discuss the problem related to the final experiment which is for model estimation; in Section 3.3 we discuss the pre-treatment of the measured data sequences. Section 3.4 contains the conclusion of the chapter.

3.1 Selection of Inputs/Outputs and Preliminary Experiments

In this book we will concentrate on the identification of linear, time-invariant, finite dimensional models of multi-input multi-output (MIMO) industrial processes. Some characteristics of industrial processes hampering this work can be listed as:

— mostly distributed parameter processes (described by partial differential equations);

— possibly nonlinear for practical ranges;

— generally time varying dynamics;

— measurement time delays involved;

— continuous-time behaviour.

We try to circumvent these obstacles by lumping techniques, compensating the nonlinear relations, restricting the ranges in amplitudes and time, and determining the delays. Therefore the following assumptions on the process are made:

— the process is operated in the neighborhood of a limited set of working points;

— the process stays near a certain working point for a long time compared to the largest time constant;

— at each working point the process dynamics are time-invariant or the change is very slow;

— at each working point the disturbances are considered to be stationary;

— at each working point the process behaviour can be described sufficiently accurate by linear models.

A real process will not always completely satisfy these assumptions, hence it is important to test the behaviour of the process with respect to the validity of the assumptions. Several of these tests will be discussed in this section. First the choice of process inputs and outputs will be discussed.

Selection of Inputs and Outputs

Usually process outputs are those variables to be controlled or to be predicted (simulated). More discussion on this can be found in Section 9.1.

Some rules for the selection of the inputs for control and simulation purposes are:

1) the selected inputs should have strong influence on (part of) the outputs both in amplitudes and in wide frequency range;

2) different inputs have essentially different effects on the outputs;

3) reliable manipulation of the inputs is possible (for control purpose);

4) input/output transfers have to be almost linear or linearizable (nonlinearities such as thresholds, dead zones, hysteresis which occur in process actuators, e.g., valves and motors, can be compensated using primary single loop controllers);

5) if an input can be measured but cannot be manipulated, use it as an input signal in the identification; the identified transfers from this input to the outputs can later on be used to design feedforward compensator (this implies that the number of the inputs for identification can be larger than that for control);

6) more manipulating inputs than the outputs to be controlled (regulated).

The use of experience of the process operators and any available *a priori* knowledge of the process are essential for the inputs selection. Some simple experiments using step inputs may also be needed for this purpose.

If the model is to be used for fault diagnosis, one also need to take care that the phenomena to be detected are included in certain channels of the selected inputs and outputs.

When appropriate candidate inputs and outputs have been chosen, experiments directed to the rough modelling of the process can be carried out. In general three types of experiments are required and we call them *preliminary experiments*.

Free-Run Experiment

In this type of experiments none of the inputs are activated. The process is

24

operating in an open loop. The intervention of the operator should be kept to the minimum. The process outputs are measured over such a long period of time that the statistical properties of the measured output signals no longer show significant change. This type of experiments are directed to finding the characteristics of the disturbances on the process outputs. The range of the amplitudes of these signals and their frequency spectra are important information for both the identification and control of the process. This experiment is purely a measurement of the existing process without introducing any disturbance to it. This is technically easy and economically cheap.

Staircase Experiment

This type of experiment is directed to testing the range of linearity of the process, to finding the static gains and to a rough estimation of the largest relevant time constant. Staircase test signals are applied to the selected candidate process inputs; see Fig. 3.1.1. The time interval of one stair should allow the process to reach its steady state.

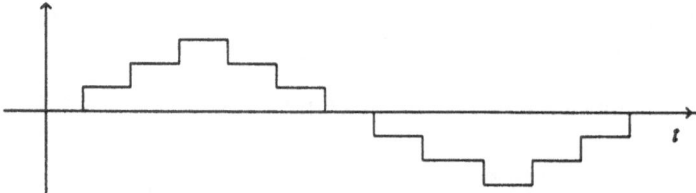

Fig. 3.1.1 Staircase test input signal

The steady state response of the process to these inputs may be used to test the steady state linearity of the process. We also can estimate the largest relevant time constant from these responses. An estimate of the largest time constant is required to determine the duration of an experiment for parameter estimation (final experiment). A rule of thumb is that the duration of the final experiment for parameter estimation should be longer than 10 times the largest time constant in order to have a reliable estimate of the model at low frequencies.

To test the static nonlinearity, a polynomial is fitted to the measured steady state responses. By solving a set of linear equations in the least-squares sense the coefficients of the polynomials can be obtained. A third order polynomial is often sufficient. If a strong nonlinearity shows up, we can either use the inverse of the nonlinear model as a precompensatore or simply do not use the related input.

It is possible that staircase experiment is not permitted for some processes due to technical and economical reasons. In this case we recommend to use steps on the inputs. Steps are mostly acceptable because they are often used by process operators for regulating the processes. The largest relevant time constant can be estimated from the step responses; however, no knowledge about the static nonlinearity can be obtained from them.

White Noise Experiment

The third set of experiments are for determining the process bandwidth and time delays. In this type of experiments, the inputs are supplied with mutually independent *white noise* signals. A discrete-time white noise $u(t)$ is a sequence of independent and identically distributed random variables of zero mean and variance R. Thus the autocorrelation function of the white noise signal is:

$$R_u(\tau) = E[u(t)u(t-\tau)] = \begin{cases} R & \text{for } \tau = 0 \\ 0 & \text{for } \tau \neq 0 \end{cases} \tag{3.1.1}$$

and its power spectrum density is constant over all frequencies. According to formula (2.4.11), we can estimate each frequency response by

$$G_{ij}(e^{i\omega}) = \frac{\hat{\Phi}_{y_i u_j}(\omega)}{\hat{\Phi}_{u_j}(\omega)} \tag{3.1.2}$$

where $\hat{\Phi}_{y_i u_j}(\omega)$ and $\hat{\Phi}_{u_j}(\omega)$ are the estimated cross-spectrum between $u_j(t)$ and $y_i(t)$ and the auto-spectrum of $u_j(t)$. Because white noises have broad frequency bands (their spectral dencities are constants over the frequency) and the disturbances usually concentrate in the low and middle frequency band, the estimates of the frequency responses are accurate at high frequencies. From these estimates the bandwidth, or equivalently the smallest time constant, of the process can be determined. The bandwidth of the open loop process is used to determine the sampling frequency of the final identifition experiment which is also the sampling frequency of computer controlled system. This frequency is called the *working frequency*.

It is known that when the inputs of a linear process are independent white noises, the cross-correlation functions between inputs and outputs are the impulse responses of the corresponding input/output transfer function. From this fact we can estimate the delays of the various input/output transfers from the computed cross-correlation functions.

The white noise signals can be approximated by PRBS (pseudo random binary sequence) with a high clock frequency. This type of signals is often used in practice because the power is constant and fixed, while also the amplitude is fixed and minimal for the given power.

The sampling frequency of the white noise experiment can be higher than the *working frequency* that is used for the modelling and control. On the other hand, if a white noise is applied to a slow process with too high a sampling frequency, there will be hardly any response from the process. So some trade-off may be required.

3.2 Experiment for Model Estimation

This experiment is directed to the collection of the input/output data that will be used to estimate the process model. This will be the final identification experiment if the whole identification procedure is later on proven to be successful. In the literature, the topics about experiment design and input design are mostly related to this final experiment. Many problems need to be treated for this type of experiment; some of them can be resolved using simple devices, like changing the sampling frequency and anti-aliasing filter; some of the problems can be theoretically very involved and numerically difficult, like optimal input design. In this chapter we aim at a good design for the final experiment such that the collected data are informative enough for the model estimation. An optimal design problem will be treated in Chapters 7 and 8.

Sampling Frequency and Anti-Aliasing Filter

In a computer-based data acquisition system, it is unavoidable that sampling leads to information losses. Therefore, it is important to select the sampling frequency so that these losses are insignificant for the given application. Industrial processes are typically slow with respect to the sampling speed of nowadays computer systems. Although it is in general true that a higher sampling frequency will cause less loss of information, the following factors may prevent us from sampling as fast as possible:

— building discrete-time models with a very small sampling time compared to the natural time constants is a numerically sensitive procedure, because all poles cluster around the point $(1, i0)$ on the z-plane;

— if the model is biased, the model fit may be concentrated in a high frequency band (Wahlberg and Ljung, 1986), which is not desirable for most of the model applications.

— a fast sampled model will often be nonminimum phase even if the original continuous-time process is minimum phase (Åström and Wittenmark, 1984), and a process with time delays may be modelled with delays of many samples, such . effects will cause difficulties in the control design.

Denote the smallest process time constant of interest to the user by τ_{min} seconds and denote T as the sampling time. Then a reasonable relation between the two is

$$T \approx \tau_{min}/3 \quad (s)$$

This implies that the sampling frequency is

$$f_s = \frac{3}{2\pi \cdot \tau_{min}} \quad (Hz) \tag{3.2.1}$$

As mentioned before, the smallest time constant can be estimated from the data of white noise experiment.

The information loss due to sampling is best described in the frequency domain. Denote $\omega_s = 2\pi f_s$ as the sampling frequency in rad/s, then the *Nyquist frequency is* $\omega_N = \omega_s/2$. It is known that the part of the signal spectrum that corresponds to frequencies above the Nyquist frequency ω_N will be interpreted as contributions from lower frequencies. This is the *alias* phenomenon or frequency folding (Åström and Wittenmark, 1984).

The information about the frequencies higher than the Nyquist frequency is thus lost by sampling. It is then important not to make bad worse by letting the aliasing effect distort the interesting part of the spectrum below the Nyquist frequency. This is achieved by a presampling filter or *anti-aliasing* filter $K(s)$ which is an analog filter. Normally the cutoff frequency of the filter is equal to the Nyquist frequency. Such filters should always be applied before sampling if we suspect that the signal has non-negligible energy above the Nyquist frequency.

Anti-aliasing filters can also be used for noise reduction. A typical situation in identification experiment is that the signal consists of a useful part and disturbance/noise part. For industrial processes, usually the disturbances have low frequency content, while measurement noises are more broadband than that of the useful signal. Then the sampling time is usually chosen so that most of the spectrum of the useful part is below ω_N. The anti-aliasing filter then cuts away the high frequency noise content.

The scheme of applying anti-aliasing filter in identification data acquisition is shown in Fig. 3.2.1; here $u(t)$ and $y(t)$ denote continuous-time input and output, and $u(kT)$ and $y(kT)$ denote the sampled input and output. Note that the filter is *applied on both input and output signals*.

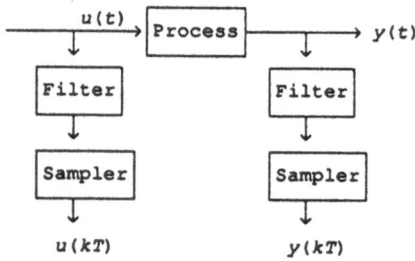

Fig. 3.2.1 Using anti-aliasing filters in data acquisition

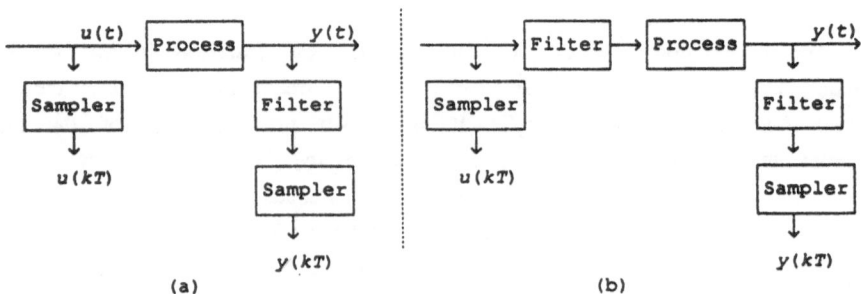

(a)   (b)

Fig. 3.2.2 Using anti-aliasing filters in the wrong places

Two schemes where the anti-aliasing filter is incorrectly placed are shown in Fig. 3.2.2. In Fig. 3.2.2 (a) the filter will be included in the identified model; in Fig. 3.2.2 (b) the filter will be included in the model twice. When the anti-aliasing filter is included in the model, it will not cause significant change of the model amplitude, however, this is not true for the phase change which is usually not negligible. This cause problems in control design.

Test Signals, Persistent Excitation, and Signal Amplitudes and Spectra

Three types of test signals are commonly used for the final experiment:

- Pseudo random binary sequence (PRBS) (see Eykhoff, 1974 or Söderström and Stoica, 1989).

- Filtered white noise or autoregressive moving average (ARMA) process.

- Sum of sinusoids.

For good description and analysis of these signal see Söderström and Stoica (1989). Here we will comment on how to use these types of signals to design a good identification experiment.

Remark— By test signals we mean the signals generated by the user and supplied to the process. In open loop experiments, the process inputs are the same as the test signals, except for the offsets. In closed loop experiments, however, the inputs are *not* the same as the test signals.

The model of the process is estimated from the input/output data collected from the final experiment. In order to guarantee that the estimation algorithms have unique solutions (the so called identifiability problem), some minimum requirement should be imposed on the test signals. This is called *persistent excitation* condition. We will introduce the concept only for single variable signals.

A discrete-time signal $u(t)$ is said to be persistent exciting of order n if the following limit exists:

$$R_u(\tau) = \lim_{N \to \infty} \frac{1}{N} \sum_{t=1}^{N} u(t)u(t-\tau) \qquad (3.2.2)$$

and the matrix

$$
\begin{bmatrix}
R_u(0) & R_u(1) \cdots R_u(n\text{-}1) \\
R_u(\text{-}1) & R_u(0) \quad \vdots \\
\vdots & \ddots \\
R_u(1\text{-}n) & \cdots \quad R_u(0)
\end{bmatrix}
\tag{3.2.3}
$$

is positive definite. This guarantees that the matrix is invertible.

The concept of persistent excitation was introduced for the estimation of impulse response model. However, the concept can also be extended to other model forms. It can be shown that a necessary condition for consistent estimation of a nth order linear process is that the test signal is persistently exciting of order $2n$ (Ljung, 1987, Söderström and Stoica, 1989).

The frequency domain interpretation of persistent excitation of order n is that the spectrum of the signal is nonzero in at least n frequencies in the interval $(-\pi, \pi)$ (Ljung, 1987, Söderström and Stoica, 1989). Based on this property we give some comments on commonly used test signals.

Sum of sinusoids. We consider the sum of n sinusoids

$$
u(t) = \sum_{j=1}^{n} A_j \sin(\omega_j t + \varphi_j) \qquad 0 < \omega_1 < \omega_2 < \cdots < \omega_n < \pi
$$

It is well known that its spectrum has exactly $2n$ lines in $(-\pi, \pi)$. Thus for the identification of an nth order process at least n sinusoids need to be injected to the process. For good noise reduction the number of sinusoids should be much greater than the process order n.

Pseudo random binary sequence (PRBS). Consider a PRBS of period M. It can be shown that the order of persistent excitation of the PRBS is M. When the clock time of the PRBS signal is small, which is the case when we want to approximate a white noise, M is rather large which is enough for most identification methods.

Filtered white noise. This type of signals has a continuous spectrum over the frequency range. Hence a filtered white noise is persistent exciting for any finite order. Thus we do not need to worry about this problem when using filtered white noises as the test signals.

When the persistent of excitation is assured, the selection of signal amplitudes and spectra still needs to be made. The amplitudes of the test signals should be chosen such that the signal-to-noise ratio is reasonably high and, at the same time, the process is not driven out of the linear range. Also the production economics and safety considerations will set constraint on the amplitudes of the test signals.

When the type and the amplitudes of the test signals are determined, it is then important to chose the proper spectral distributions of the test signals. Often the so called *optimal input design* is referred to the design of test input spectrum (Ljung, 1987). Logically, the signal spectra should be designed in such a way that the identified model is *best for the intended use of the model*. Therefore, the test signal design will be influenced by model application. A general guideline for the spectra design is that the experimental condition should resemble the situation for which the model is going to be used.

If the model is used for simulation purpose, the spectrum of the test input should be similar to that of the simulation input. In Chapter 7 we shall see that the spectrum of optimal test input is a function of the spectrum of the simulation input and that of the disturbance.

If the model is used for controller design in order to reduce the sensitivity to disturbances, the test input should resemble the character of the disturbance; see Chapter 7.

If the model is used for fault diagnosis, the spectrum of the test input should have such a distribution so that the parameters related to the faults are identified accurately; see Chapter 10.

When the amplitudes and spectra of the test signals are determined, they can be realized by filtering white noises. These realized signals are then used to activate the process inputs. Note that for a MIMO process, *different inputs are excited simultaneously.*

If it is not costly, it is advisable to use a sampling frequency that is a multiple of the desired working frequency. This will build redundancy into the measured data that can reduce the information loss in the pre-treatment of the data which will be discussed in the next section.

3.3 Pre-treatment of Data

When the data have been collected from the final experiment, they are often not suitable for immediate use in identification algorithms. There are several possible deficiencies:

— High frequencies disturbances in the data;

— occasional spikes and outliers;

— drift and offset, low-frequency disturbance;

— the numerical values of different inputs and outputs are not in the same order of magnitude.

When the data are collected, we must always first plot the data in order to inspect them for these deficiencies. In this section we shall discuss how to repair the data so as to avoid problems at the estimation steps later.

Peak Shaving

Peak shaving is required to reduce the effects of spikes (peaks due to e.g. loose contacts, power failure, cross-talk between cables etc.) on measured process signals. In industrial situations spikes are often induced in the sensors and the long leads from the sensors to the measuring equipment.

The amplitudes of the noise spikes may be very large compared to the actual signal range. As the energy contents of the spikes may be large, their potential influence on the estimation results can be considerable. Removal of spikes from the process signals can be accomplished using the following signal processing procedure (Backx and Damen, 1989):

— Clip the signal amplitudes to values never to be reached by the real process signals (using *a proiri* knowledge);

— compute the standard deviation of the clipped signal;

— interpolate all samples of the original signal that are outside a band defined by the trend signal $\pm \alpha$ times the computed standard deviation, where α is chosen such that no signal value exceeds the permitted signal range.

All consecutive sample values outside the permitted band are replaced by values obtained from a linear interpolation starting from the last accepted sample value and stopping at the first sample value within the permitted band after the spike. If slow variations accur in the data a trend correction need to be performed before the peak shaving operation.

Trend Correction

Drifts or slow variations are not uncommon in industrial data. The causes of them are in most cases known. Usually it is not possible to prevent drifts and slow variations during open loop experiments. Trends and drifts in general have bad influence on the estimation results. They do not average out because of their low frequency behaviour and they will cause considerable bias of the model. On the other hand, this type of disturbance can easily be compensated for by the feedback control. Hence if the model is used for control design, there is no need to model this part of the signals.

We recommend two approaches to trend correction. The first one is to estimate the trends and remove them from the signals. A good approach for the computation of the trend of a signal is to use a symmetric, noncausal, lowpass filter, as such filter has no phase shift (Backx, 1987 and Backx and Damen, 1989). This filter can be accomplished off-line by filtering the signal with a causal lowpass filter of appropriate cut-off frequency both in forward and reversed time.

The second method of trend correction is to perform bandpass filtering on inputs and outputs. The band of the filter should cover the dynamics of the process we are interested in. In this way we can remove the trend/drifts and high frequency noise from the data at once. Note that this method does not supply a tend of the given signal.

Scaling and Offset Correction

In industrial practice not all inputs and outputs have the same order of magnitude. The numerical values obtained are related to physical quantities which, in general, do not have the same dimensions. The signals with the largest numerical values will automatically get highest weight in the quadratic loss function which is minimized for determining the model, if these signals are directly used. This problem may be overcome by correcting the signals for offsets and by scaling them afterwards. The steps are:

> Subtract average signal values in order to allow the use of a linear model without offset to describe the dynamic behaviour of the process around its working point.

— Scale both input and output signals with respect to their power contents after offset correction.

Delay Correction

In the preparation for the parameter estimation, delay times can be compensated by shifting input and output signal relative to each other. One of the signals is used as a reference and the remaining signals are shifted in time in order to compensate the delays. For a process with m inputs and p outputs, in general, each output can show a delay in the response to each input so that finally $m \cdot p$ different time delays may be found. The number of delays that can be compensated by shifting the signals against each other is only $m + p - 1$, because only $m + p$ signals are available and one signal is fixed as the reference. As a result not all delays can be compensated exactly. The compensation has to be such that the total sum of delays left is minimal. Time delays remaining after the compensation have to be estimated as part of the model.

Filtering and Sampling Rate Reduction

Remember that we have built redundancy into the data by sampling faster than necessary. If there is noise energy in the high frequency band beyond the interested range, we can use digital low pass filtering to cut it off. Do filter *all* the inputs and outputs. Suppose that the sampling frequency is l times the working frequency, we reduce the frequency to the working frequency by taking one sample from each l samples.

Now the data can be used for model estimation. Before an estimated model is used, some validation of the model should be carried out. Usually we need another set of data for model validation. This set can be generated on another occasion; or we can simply cut the data from the same experiment into two sets and use one set for estimation and the other for validation.

3.4 Conclusions

In an identification application the most crucial step is to excite good information from the process about its dynamics. This is the problem of experiment design. In this chapter we recommended four types of experiments:

— Free-run experiment;

— staircase (or step) experiment;

— white noise experiment; and

— final experiment.

The first three types are used to collect *a priori* knowledge of the process and of the disturbances. The last experiment is for model estimation.

In order to let identification algorithms work, or to guarantee the identifiability, the test signals of the final experiment should be persistently exciting with certain order. If we want to do better than that, the spectra of the test signals should have proper distributions over the frequencies. This is related to the use of the model. Before using the data for model estimation, do not forget to polish them by peak shaving, trend correction, offset correction, scaling and delay compensation.

If each type of experiments is performed properly, it will be most likely that the identification will have a successful outcome. Indeed, all the four types of experiments have been performed for the three industrial processes which are used in this book to illustrate different identification methods; see Chapters 4, 5, 6, 8 and 9. When identification experiments are too costly, we may skip the staircase and white noise experiments and only perform the free-run experiment and the final experiment. In the free run experiment, no excitation is introduced to the process, in fact we are only measuring the existing process. This type of experiment is the cheapest compared to the other types. The final experiment is absolutely necessary, because without a set of informative data of the process, identification will become meaningless. In Chapter 7 we will give a method of (nearly) optimal test input design for control application when only free-run and final experiments are permitted.

Often performing the experiments in a closed loop with feedback control can decrease the cost and increase the safety. All the guidelines discussed in this chapter will still hold for closed loop experiments. In closed loop case, the persistent excitation should be referred to the test signals and not to the process inputs. Because of the feedback, the persistent excitation of process inputs along cannot guarantee the identifiability of the process (Ljung, 1987).

If the purpose of identification is control, we can view the closed loop system as the "process", perform the identification experiments and estimate a model of the system. Then design a second loop controller based on the model of the closed loop system. The model of the open loop process is not necessary in this scheme. Fig. 3.4.1 shows the idea of the solution which can be called *two-loop scheme*. In doing so, we avoid the problem of closed loop identification which is more complex than the open loop problem. This *scheme* was proposed by Zhu (1990) and tested on a servo-process which is open loop instable; see also Zhu et.al. (1990).

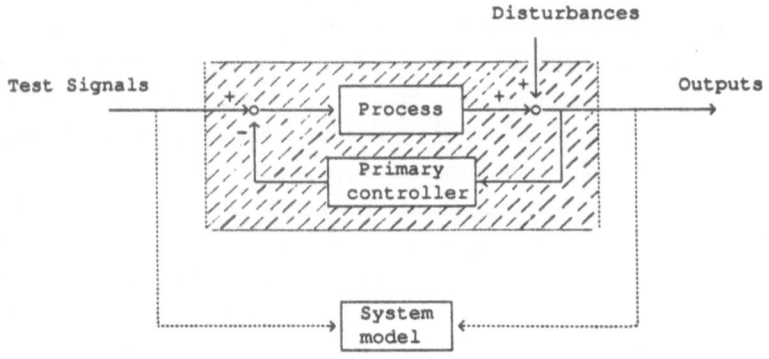

A)   Identification of the closed loop system

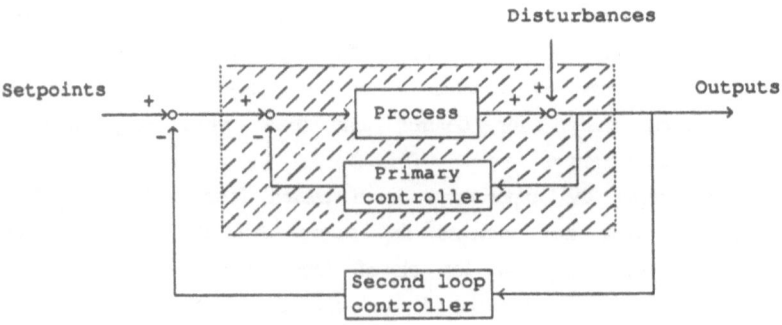

B)   Design of the second loop controller

Fig. 3.4.1   The two-loop scheme for combining identification and control

CHAPTER 4

IDENTIFICATION BY THE LEAST-SQUARES METHOD

In this chapter we will study the least-squares method. The least-squares principle was invented by Karl Gauss at the end of the eighteenth century for determining the orbits of planets. Since then this method has become a major tool for parameter estimation using experimental data. Most existing parametric identification methods can be related to the least-squares method. The method is easy to comprehend and, due to the existence of a closed solution, it is also easy to implement. The least-squares method is also called *linear regression* (in statistical literature) and *equation error method* (in identification literature).

We will first introduce the principle (Section 4.1); then apply it for the estimation of finite impulse response (FIR) models and the estimation of parametric models (Section 4.2). Before making assumptions and pursuing a theoretical analysis of the method, we first test the method on two industrial processes, a single stand rolling mill and a glass tube production process (Section 4.3) where also the method is extended to a multivariable case. The method is successful for the first process; but it fails for the second one. Why? In Section 4.4 some theoretical analysis is carried out, and reasons are given why the least-squares method can fail. Finally in Section 4.5 we will draw conclusions on the least-squares method.

For reasons given in Chapter 2, in most of the chapters discrete-time models are treated; in Chapter 10 we study continuous-time model estimation.

4.1 The Principle of Least-Squares

The least-squares technique is a mathematical procedure by which the unknown parameters of a mathematical model are chosen (estimated) such that the sum of the squares of some chosen error is minimized. Suppose a mathematical model is given in the form

$$y(t) = x_1(t)\theta_1 + x_2(t)\theta_2 + \cdots + x_n(t)\theta_n \qquad (4.1.1)$$

where $y(t)$ is the observed variable, $\{\theta_1, \theta_2, \cdots, \theta_n\}$ is a set of constant parameters, $x_1(t)$, $x_2(t)$, $\cdots$, $x_n(t)$ are known functions that may depend on other known variables. The variable t often denotes time.

Assume that N samples of measurements of $y(t)$ and $x_1(t)$, $x_2(t)$, $\cdots$, $x_n(t)$ are made at time 1, 2, $\cdots$, N. Filling the data samples into equation (4.1.1) results in a set of linear equations

$$y(t) = x_1(t)\theta_1 + x_2(t)\theta_2 + \cdots + x_n(t)\theta_n \qquad t = 1, 2, \cdots, N \qquad (4.1.2)$$

This can be arranged in a simple matrix form

$$y = \Phi\theta \qquad (4.1.3)$$

where

$$y = \begin{bmatrix} y(1) \\ y(2) \\ \vdots \\ y(N) \end{bmatrix}, \quad \Phi = \begin{bmatrix} x_1(1) \; x_2(1) \; \cdots \; x_n(1) \\ x_1(2) \; x_2(2) \quad\;\; x_n(2) \\ \vdots \quad\; \vdots \quad \ddots \quad \vdots \\ x_1(N) \; x_2(N) \; \cdots \; x_n(N) \end{bmatrix}, \quad \theta = \begin{bmatrix} \theta_1 \\ \theta_2 \\ \vdots \\ \theta_n \end{bmatrix}$$

A necessary condition that this set of equations has a solution is $N \geq n$. For $N = n$, we have a unique solution

$$\hat{\theta} = \Phi^{-1}y \qquad (4.1.4)$$

provided Φ^{-1}, the inverse of the square matrix Φ, exists. $\hat{\theta}$ denotes the estimate of θ. This is well known. However, when $N > n$, it is generally not possible to find a $\hat{\theta}$ vector which can fit the data samples exactly, because the data may be contaminated by disturbances and measurement noise. A wrong model order or a wrong model structure are other possible causes of misfit. A way to determine the parameters is to estimate them on the basis of least-squares-error.

Introduce a residual (error), $\varepsilon(t)$, and let

$$\varepsilon(t) = y(t) - \hat{y}(t) = y(t) - \varphi(t)\theta$$

Now we will choose $\hat{\theta}$ such that the criterion (loss function)

$$V_{LS} = \frac{1}{N}\sum_{t=1}^{N}\varepsilon(t)^2 = \frac{1}{N}\sum_{t=1}^{N}[y(t) - \varphi(t)\theta]^2 = \frac{1}{N}\underline{\varepsilon}^T\underline{\varepsilon} \qquad (4.1.5)$$

is minimized. Here

$$\underline{\varepsilon} = \begin{bmatrix} \varepsilon(1) \\ \varepsilon(2) \\ \vdots \\ \varepsilon(N) \end{bmatrix}$$

To carry out the minimization, we express

$$V_{LS}(\theta) = \frac{1}{N}(y - \Phi\theta)^T(y - \Phi\theta) = \frac{1}{N}[y^Ty - \theta^T\Phi^Ty - y^T\Phi\theta + \theta^T\Phi^T\Phi\theta]$$

Taking the first derivative of $V_{LS}(\theta)$ with respect to θ and equating the result to zero, we have

$$\left. \frac{\partial V_{LS}(\theta)}{\partial\theta} \right|_{\theta=\hat{\theta}} = \frac{1}{N}[-2\Phi^Ty + 2\Phi^T\Phi\hat{\theta}] = 0$$

Hence the solution is given by the equation

$$\Phi^T\Phi\hat{\theta} = \Phi^Ty \tag{4.1.6}$$

or

$$\hat{\theta} = [\Phi^T\Phi]^{-1}\Phi^Ty \tag{4.1.7}$$

This result is the well known *least-squares estimator* of θ.

From linear algebra, we know that equation (4.1.6) will have an unique solution if and only if the matrix

$$\Phi^T\Phi = \sum_{t=1}^{N} \varphi^T(t)\varphi(t) \tag{4.1.8}$$

is nonsingular. This is called the parameter identifiability condition which has been discussed in Chapter 3.

The criterion (4.1.5) can be generalized by introducing a weighting matrix to allow each error term to be weighted individually. Let W be the desired weighting matrix which is symmetrical positive definite, then the weighted error criterion becomes

$$V_w(\theta) = \frac{1}{N}\underline{\varepsilon}^TW\underline{\varepsilon} = \frac{1}{N}(y - \Phi\theta)^TW(y - \Phi\theta)$$

Minimization of $V_w(\theta)$ with respect to θ follows the same procedure as above, the parameter estimator is given by

$$\hat{\theta}_w = [\Phi^T W \Phi]^{-1} \Phi^T W y \tag{4.1.9}$$

This is called *weighted least-squares estimator*. If $W = I$, this reduces to the normal least-squares estimator.

4.2 Estimating Models of Linear Processes

Now we will show how the least-squares method can be used to estimate models of linear dynamic processes. The particular way to do this will depend on the character of the model and its parametrization.

4.2.1 Finite Impulse Response (FIR) Models

A linear time-invariant dynamic process is uniquely characterized by its impulse response. For stable processes the impulse response will tend to zero for increasing time and may then be truncated. This results in so called finite impulse response (FIR) models or Markov parameter models. For single-input single-output processes such a model is given by

$$y(t) = g_1 u(t\text{-}1) + g_2 u(t\text{-}2) + \cdots + g_n u(t\text{-}n)$$

$$= \sum_{k=1}^{n} g_k u(t\text{-}k) = \varphi(t)\theta \tag{4.2.1}$$

where

$$\theta = \begin{bmatrix} g_1 \\ g_2 \\ \vdots \\ g_n \end{bmatrix}, \quad \varphi(t) = [u(t\text{-}1) \ u(t\text{-}2) \ \cdots \ u(t\text{-}n)]$$

This model is identical to the regression model in (4.1.1) when replacing $x_k(t)$ by $u(t\text{-}k)$. Hence the least-squares estimator (4.1.7) can be applied directly.

Given the measured input/output data sequence:

$$y(1), \ u(1), \ \cdots \cdots, \ y(N+n), \ u(N+n)$$

and introducing the residual as

$$y(t) = g_1 u(t-1) + g_2 u(t-2) + \cdots + g_n u(t-n) + \varepsilon(t)$$
$$= \varphi(t)\theta + \varepsilon(t)$$

then we have

$$\hat{\theta} = [\Phi^T \Phi]^{-1} \Phi^T y \qquad\qquad (4.2.2)$$

where

$$\Phi = \begin{bmatrix} u(n) & u(n-1) \cdots & u(1) \\ u(n+1) & \cdot & \vdots \\ \vdots & \vdots & \vdots \\ u(n+N) & \cdots & \cdots & u(N+1) \end{bmatrix}$$

Fig. 4.2.1 shows the block diagram of FIR model estimation.

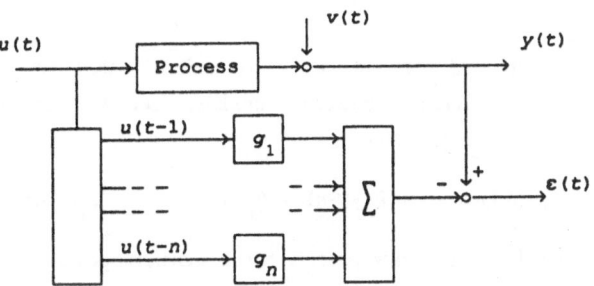

Fig. 4.2.1   Error generation of FIR model estimation

The advantages of FIR models are the following:

1)   Less *a priori* knowledge of the process is required to estimate a FIR model. The troublesome questions of the order or structure of the process as required when using other models do not arise.

2)   FIR model estimate is statistically unbiased (the expectation of the estimate equals the true value) and consistent (the estimate tends to the true value when the number of data points tends to infinity); see Section 4.4 for the analysis.

Of course FIR models also have disadvantages:

1)   This model structure often needs many parameters; the estimated process transfer function is not accurate for finite number of data, because the variance of the estimate is proportional to the number of parameters of the FIR model (see Chapter 7).

2)   FIR model is not suitable for most of the existing control design methods. Often a model reduction is necessary to arrive at a compact parametric model; see Chapter 6.

The extension of the FIR model estimation to multivariable processes is straightforward. We will show this by a case study in the next section.

4.2.2   Transfer Operator Models

The least squares method can be used to estimate the parameters of transfer function models or difference equation models. Let the process be described by an nth order difference equation

$$y(t) + a_1 y(t-1) + \cdots + a_n y(t-n) = b_1 u(t-1) + \cdots + b_n u(t-n) \qquad (4.2.3)$$

From Chapter 2 we know that this process has a transfer operator defined by

$$G^\circ(q) = \frac{B(q)}{A(q)} \qquad (4.2.4)$$

where

$$A(q) = 1 + a_1 q^{-1} + \cdots + a_n q^{-n}$$
$$B(q) = \qquad b_1 q^{-1} + \cdots + b_n q^{-n}$$

Given the input/output data sequence as

$$y(1)\ u(1)\ \cdots\ \cdots\ y(N+n)\ u(N+n)$$

and assuming that the order n is known, then we need to estimate the parameters a_i and b_i. To do this, we introduce the error as

$$y(t) + a_1 y(t-1) + \cdots + a_n y(t-n) = b_1 u(t-1) + \cdots + b_n u(t-n) + \varepsilon(t) \qquad (4.2.5)$$

or

$$A(q)y(t) = B(q)u(t) + \varepsilon(t)$$

The term $\varepsilon(t)$ is used to account for the fitting error. In literature $\varepsilon(t)$ is referred to as the *residual* or *equation error*.

Rewrite equation (4.2.5) as

$$y(t) = -a_1y(t-1) - \cdots - a_ny(t-n) + b_1u(t-1) + \cdots + b_nu(t-n) + \varepsilon(t)$$
$$= \varphi(t)\theta + \varepsilon(t) \tag{4.2.6}$$

where $\varphi(t)$ is the data vector

$$\varphi(t) = [-y(t-1), \cdots -y(t-n) \quad u(t-1) \cdots u(t-n)]$$

and θ is the parameter vector

$$\theta = \begin{bmatrix} a_1 \\ \vdots \\ a_n \\ b_1 \\ \vdots \\ b_n \end{bmatrix}$$

Using the data sequence we can form a system of N equations ($N \gg 2n$):

$$y = \Phi\theta + \underline{\varepsilon} \tag{4.2.7}$$

where

$$y = \begin{bmatrix} y(n+1) \\ y(n+2) \\ \vdots \\ y(n+N) \end{bmatrix}, \quad \underline{\varepsilon} = \begin{bmatrix} \varepsilon(n+1) \\ \varepsilon(n+2) \\ \vdots \\ \varepsilon(n+N) \end{bmatrix}$$

$$\Phi = \begin{bmatrix} \varphi(n+1) \\ \varphi(n+2) \\ \vdots \\ \varphi(n+N) \end{bmatrix} = \begin{bmatrix} -y(n) & \cdots & -y(1) & | & u(n) & \cdots & u(1) \\ -y(n+1) & & \vdots & | & u(n+1) & & \vdots \\ \vdots & & \vdots & | & \vdots & & \vdots \\ -y(n+N) & \cdots & -y(N+1) & | & u(n+N) & \cdots & u(N+1) \end{bmatrix}$$

Then according to the least-squares principle of Section 4.1, the estimate which minimizes the loss function

$$V_{LS} = \sum_{t=n+1}^{N+n} \varepsilon(t)^2 = \sum_{t=n+1}^{N+n} [y(t) - \varphi(t)\theta]^2 \tag{4.2.8}$$

is

$$\hat{\theta} = [\Phi^T\Phi]^{-1}\Phi^T y = \left[\sum_{t=n+1}^{N+n} \varphi(t)^T\varphi(t) \right]^{-1} \left[\sum_{t=n+1}^{N+n} \varphi(t)^T y(t) \right] \tag{4.2.9}$$

This solution exists if matrix

$$\Phi^T\Phi = \sum_{t=n+1}^{N+n} \varphi(t)^T\varphi(t)$$

is nonsingular, which can be guaranteed if the process order is n and the input signal $u(t)$ is persistently exciting with order $2n$ (see Chapter 3). Fig. 4.2.2 shows the block diagram of error generation in transfer operator estimation by the least-squares method.

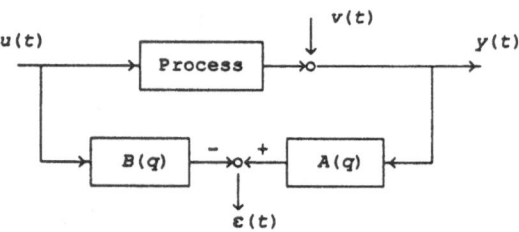

Fig. 4.2.2   Error generation in transfer operator estimation

The advantage of parametric identification by the least-squares method, or *equation error method*, is its numerical simplicity. The method has a closed solution. This is due to the fact that the error (residual) is linear in parameters a_i and b_i and that a quadratic loss function (4.2.8) is minimized. It will work well when the noise level is low and model order is correct. Under practical conditions, however, the least-squares estimator of the transfer function model is biased; see Section 4.4. This will cause accuracy problems if the noise level increases. Moreover, when estimating a transfer function (frequency response), the least-squares estimator can give a poor fit to the process transfer function at low and middle frequencies; see Section 4.4. This is of course not desired in applications such as control and simulation.

The extension of the method to multivariable processes is straight-forward if diagonal form MFD models (see Chapter 2) are used. This will be shown by a case study in Section 4.3.

4.2.3 Model Validation and Order Selection; A Simulation Approach

The identification method discussed so far assumed that the order of the model is known, and that only the parameters are to be estimated. This is only part of the story because, in practice, the order of the process is seldom exactly known. Therefore model order or structure determination is an important topic in process identification. The literature on order or structure selection is enormous. At this stage of development, we present a simple and practical technique. Model order selection is closely related to model validation which means to check whether a model is good enough for the intended use of the model. In practice one can perform identification for increasing model orders, and stop when a model is satisfactory using some model validation method.

We will revisit these problems in the following chapters when we know more about identification.

For model applications such as control and simulation, a simple but very effective method of model validation is to simulate the estimated model using some test input, and compare the model output with the measured output; see Fig. 4.2.3.

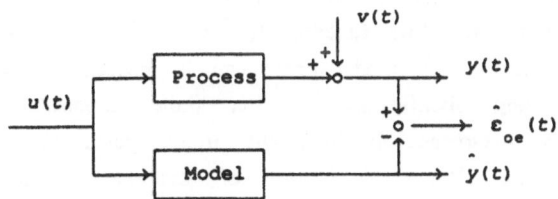

Fig. 4.2.3 Model validation by simulation

The goodness of fit is measured by the sum of the squares of the simulation error or *output error*:

$$V_{oe} = \frac{1}{N} \sum_{t=1}^{N} \hat{\varepsilon}_{oe}(t)^2 \qquad (4.2.10)$$

where

$$\hat{\varepsilon}_{oe}(t) = y(t) - \frac{\hat{B}(q^{-1})}{\hat{A}(q^{-1})} u(t) = y(t) - \hat{y}(t) \qquad (4.2.11)$$

We note that this error is *not* the same as the equation error introduced in least-squares method; compare Fig. 4.2.2. The relation between the two errors is

$$\hat{\varepsilon}(t) = \hat{A}(q)\hat{\varepsilon}_{oe}(t) \qquad (4.2.12)$$

We see that *when using a finite impulse response (FIR) model, then A(q) = 1 and the equation error is equal to the output error.*

For model validation we use the output error instead of the equation error because $\hat{\varepsilon}_{oe}(t)$ is the simulation error plus the disturbance, and it is more directly related to the error of the transfer function. From Fig. 4.2.3 and equation (4.2.11) we have

$$\hat{\varepsilon}_{oe}(t) = [G^{o}(q) - \hat{G}(q)]u(t) + v(t) \qquad (4.2.13)$$

Thus this error consists of the model misfit and the output disturbance.

In general, the loss function V_{oe} decreases as the order n increases. The reduction of V_{oe} ceases to be significant when the order of the model is high enough for simulation purposes. Based on this principle, a procedure for order selection is simply to compute the least-squares estimates and corresponding output error loss function for a sequence of model orders $n = 1, 2, 3, \cdots$. The appropriate model order can be chosen as the one for which V_{oe} stops decreasing significantly. For a finite dimensional process, when the level of the disturbance is low, the true process order may be found using this procedure. When the level of disturbance increases, the order determined by this procedure can be higher than the true order; the higher order model is needed to model the disturbance in order to overcome the bias of the least-squares estimator (see Section 4.4).

So far our aim is to find the best order or the true order. In industrial applications, it is often desirable to have a model which is as simple as possible, yet accurate enough for control or simulation purposes. To this end we can determine the order by using the relative output error (simulation error) loss function defined as

$$RV_{oe} = \frac{1}{N} \sum_{t=1}^{N} \varepsilon_{oe}(t)^2 \Big/ \frac{1}{N} \sum_{t=1}^{N} y(t)^2 \qquad (4.2.14)$$

Then the rule is to choose the lowest order such that this error is smaller than some given threshold. In our experience, for a large class of industrial processes, a proper threshold is between 10% and 20%, for purposes such as feedback controller design and simulation.

An important remark is that model validation and order selection should be performed on a data set which has not been used for model estimation. This is sometimes called *cross-validation*. The reason is that a good fit to the data which were used to estimate the parameters is not sufficient to prove the model quality. In the extreme case, when the model order equals the number of equations $n = N$, then the fitting error is zero; the model quality, however, can be very poor due to large variances.

Example 4.2.1 Given the process (Åström and Eykhoff, 1971)

$$G^\circ(q) = \frac{q^{-1} + 0.5q^{-2}}{1 - 1.5q^{-1} + 0.7q^{-2}} \qquad (4.2.15)$$

This process is simulated using a white noise sequence with zero mean and variance 1 as the input $u(t)$. As a disturbance $v(t)$, a white noise of zero mean is added to the output. So, the input/output data are generated by

$$y(t) = G^\circ(q)u(t) + v(t)$$

The noise-to-signal ratio at the output (in power) is

$$N/S = var(v(t))/var(y_o(t)) = 0.1, 1, 10, 100\%$$

1000 samples are generated where the first 500 are used for model estimation and the second 500 for model validation and order selection.

The variations of the loss function V_{oe} as a function of model order for different noise-to-signal ratios are plotted in Fig. 4.2.4.

We find that if the noise level is very low ($N/S = 0.1\%$), the true order can be found. As the noise level increases, one tends to choose a higher order model when using the least-error method. This is due to the bias of least squares estimate, not due to the order selection rule.

48

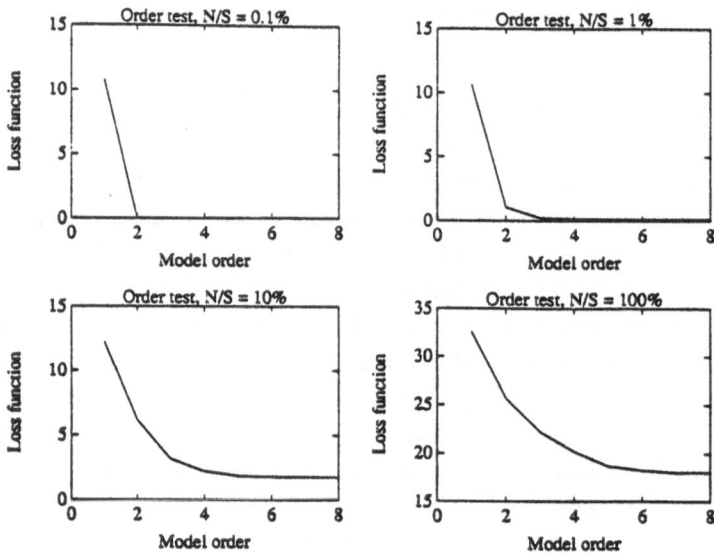

Fig. 4.2.4 Order test for different noise-to-signal ratios

Often it is not desirable to use a model order which is higher than the true order, because a high order model will need more computing power and it may cause numerical problems in controller design.

At this moment a critical reader might ask why not *estimate the parameters by minimizing the output error loss function V_{oe} directly*, which is closer to model application? The answer is simple but surprising: the least-squares method is often used because its solution is numerically simple and reliable. In fact the least-squares (equation error) loss function (4.2.8) is only a way of mathematical approximation; it is *not* a physically sensible criterion. The output error loss function is a better criterion if one wishes to identify the process transfer function. However, the output error is nonlinear in the parameters of the $A(q)$ polynomial, which means that no analytical solution for the minimization of the output error loss function (4.2.11) exists. Thus a nonlinear minimization algorithm needs to be used for finding a minimum. This is much more time-consuming than the least-squares method, and numerical problems such as local minimum and non-convergence often occur.

"We use it because it is simple" is often a way of engineering life.

Indeed, it is always advisable to first try the simplest least-squares method. In model validation and order or structure selection, however, one should not forget the purpose of modelling and use a criterion (loss function) that is as close as possible to model application; otherwise the result can be misleading. The output error criterion is a good candidate to be used.

4.3   Two Industrial Case Studies

Now we have learnt the least-squares principle and its applications in linear dynamic process identification. How does it perform? To answer the question, typically, a theorist will start making assumptions and then analyse the properties of the method; an engineer, on the other hand, may try this tool on his process and give his judgment based on test results. Obviously the two ways of life are complementary. Unlike most of the books on identification, we will first take the engineers' approach. In this section, two industrial processes are used to illustrate the least-squares method. The first process is a multi-input and single-output (MISO) process; and the second one is a two-input two-output process. Therefore the extension of the least-squares method to multi-input multi-output (MIMO) process is discussed. Also a control system design scheme will be presented.

4.3.1   Identification and Control of a Single-Stand Rolling Mill

The schematic overview of the process is shown in Fig. 4.3.1. Metal strip is rolled in order to reduce the thickness of the strip. The variation in the exit thickness is an important quality measure. Before our investigation, the rolling mill was operated manually. The increasing demand on the quality requires the reduction of the thickness variation, and this could no longer be accomplished by the operator. To reach this goal, a computer control system needs to be developed.

The exit thickness is affected by at least the following four measurable variables: drawing force of the electrical motor, roll gap, rolling speed and input thickness. The first three can be used as control variables, while the input thickness is a measurable disturbance and its variation is an important cause of the variation of exit thickness. Other causes of thickness variation are strip hardness variation and eccentricities of various rolls. These are not measurable and thus will be considered as unmeasurable disturbances.

It was decided to use drawing force and roll gap as control inputs and keep the rolling speed constant; the input thickness cannot be manipulated and will be used for feedforward control. So for the feedforward and feedback control of the process, a 3-input 1-output model needs to be identified; see Fig. 4.3.2.

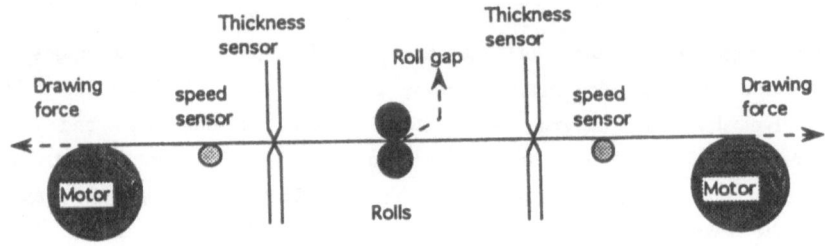

Fig. 4.3.1   The single stand rolling mill

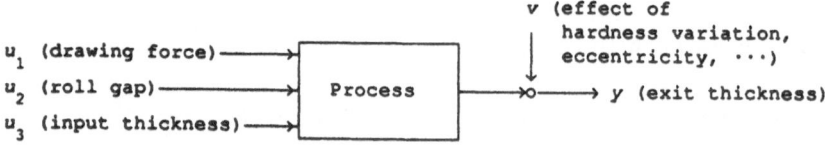

Fig. 4.3.2   A 3-input 1-output rolling mill process model

Thus the process model has the following form

$$y(t) = G_1(q)u_1(t) + G_2(q)u_2(t) + G_3(q)u_3(t) + v(t) \qquad (4.3.1)$$

Based on a staircase experiment (see Chapter 3), we learn that the process is reasonably linear around working points and the impulse responses are very short. Hence a FIR model is used in identification:

$$y(t) = \sum_{k=1}^{n_1} g^\circ_{1,k} u_1(t-k) + \sum_{k=1}^{n_2} g^\circ_{2,k} u_2(t-k) + \sum_{k=1}^{n_3} g^\circ_{3,k} u_3(t-k) + v(t) \qquad (4.3.2)$$

Introduce the residuals

$$\varepsilon(t) = y(t) - \left[\sum_{k=1}^{n_1} g_{1,k} u_1(t-k) + \sum_{k=1}^{n_2} g_{2,k} u_2(t-k) + \sum_{k=1}^{n_3} g_{3,k} u_3(t-k) \right] \qquad (4.3.3)$$

Then we are ready to apply the least-squares method.

Remark— There is a principal difference between the output disturbance $v(t)$ and the residual $\varepsilon(t)$, although they appear at the same place of equations (4.3.2) and (4.3.3). The term $v(t)$ accounts for the effect of all the unmeasurable disturbances acting at the process output; the term $\varepsilon(t)$ is used to account for model misfit which is a function of model parameters, $\varepsilon(t) = \varepsilon(t,\theta)$. Note that in equation (4.3.2) the parameters are fixed (unknown) true values; in equation (4.3.3) the parameters are the variables to be estimated.

For the model estimation input/output data are collected, where u_1 (drawing force) and u_2 (roll gap) are driven by two PRBS test signals, and u_3 (input thickness) is only measured. Denote the data sequence as

$$z^N := y(1)\ u_1(1)\ u_2(1)\ u_3(1)\ \cdots\ \cdots\ y(N+n)\ u_1(N+n)\ u_2(N+n)\ u_3(N+n)$$

where $n = \max\{n_1, n_2, n_3\}$. Then it is straightforward to extend the least-squares formula (4.2.2) for a 3-input 1-output FIR model as follows

$$\hat{\theta} = [\Phi^T \Phi]^{-1} \Phi^T \underline{y} \qquad (4.3.4)$$

where

$$\underline{y} = \begin{bmatrix} y(n+1) \\ y(n+2) \\ \vdots \\ y(n+N) \end{bmatrix}, \quad \theta = \begin{bmatrix} g_{1,1} \\ \vdots \\ g_{1,n_1} \\ g_{2,1} \\ \vdots \\ g_{2,n_2} \\ g_{3,1} \\ \vdots \\ g_{3,n_3} \end{bmatrix}$$

$$\Phi = \begin{bmatrix} u_1(n_1) & \cdots & u_1(1) & | & u_2(n_2) & \cdots & u_2(1) & | & u_3(n_3) & \cdots & u_3(1) \\ u_1(n_1+1) & & \vdots & | & u_2(n_2+1) & & \vdots & | & u_3(n_3+1) & & \vdots \\ \vdots & & \vdots & | & \vdots & & \vdots & | & \vdots & & \vdots \\ u_1(n_1+N) & \cdots & u_1(N+1) & | & u_2(n_2+N) & \cdots & u_2(N+1) & | & u_3(n_3+N) & \cdots & u_3(N+1) \end{bmatrix}$$

After the pretreatment of the data (see Chapter 3), a FIR model is estimated. Then the model is validated using a data sequence from another PRBS experiment. Fig. 4.3.3 shows part of the measured output (exit thickness) and simulated output. The relative simulation error (see (4.2.14)) is about 5%. For control purposes this is an accurate model.

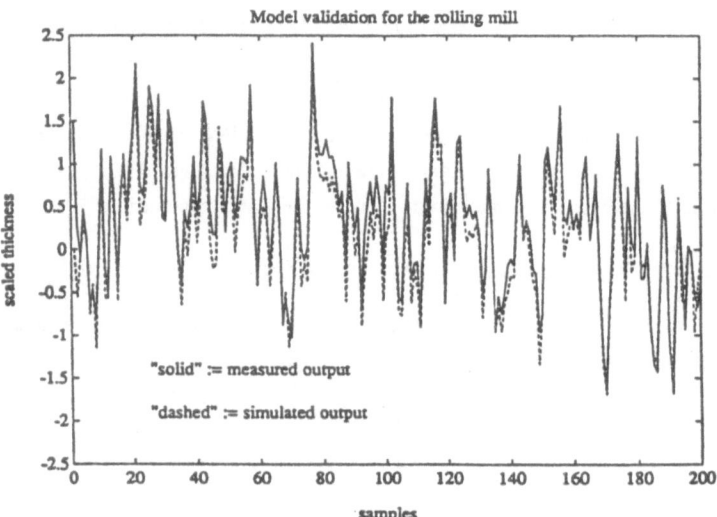

Fig. 4.3.3   Model validation for the rolling mill

Feedback and feedforward controllers have been designed based on the identified model, and on extensive simulations of the control system. Fig. 4.3.4 shows the control scheme. There exist a large measurement delay at the output; see Fig. 4.3.1. In such a case the feedback controller is only effective at low frequencies, which means that only slow disturbances can be compensated by the feedback loop. Thus, if possible, a feedforward controller should to be used to compensate for fast disturbances.

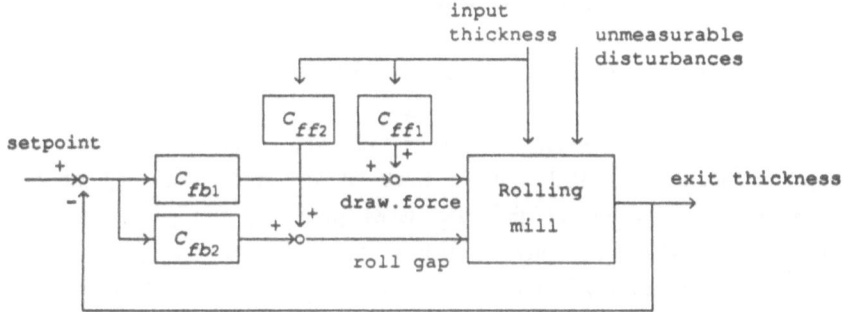

Fig. 4.3.4   The control scheme for the rolling mill, C_{fb1} and C_{fb2} are the two feedback controllers, C_{ff1} and C_{ff2} are the two feedforward controllers.

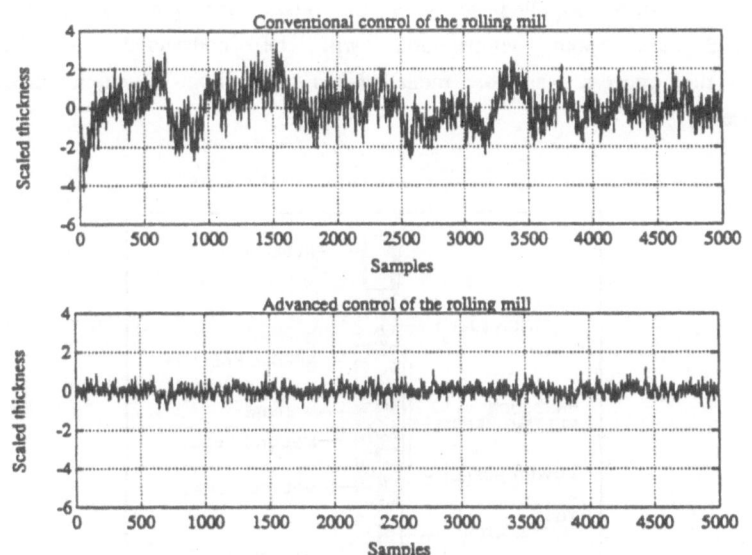

Fig. 4.3.5 Manual control versus computer control of the rolling mill

According to simulation of the controlled system, a considerable reduction of thickness variation can be achieved by the designed controllers. The industrial tests have confirmed this prediction. Fig. 4.3.5 shows the measured strip thickness during manual control and during computer control respectively. A 70% reduction of standard deviation has been achieved! This has been considered as a great achievement.

4.3.2   Identification and Control of a Glass Tube Process

The outline of the process is shown in Fig. 4.3.6. By indirect electric heating the glass is melted and it flows down through a ring shaped hole along a mandril. Shaping of the tube takes place at, and just below, the end of the mandril. The glass tube is pulled down due to gravity and a drawing machine. Two measures of dimension, wall thickness and diameter of the tube, are the most important quantities to be regulated; hence they are the outputs of the process. Many different types of tubes (differing in dimensions) are produced in the same process. The mandril gas pressure and the drawing speed can affect the wall thickness and diameter most directly and easily; so they are good candidates for control inputs. Other variables, such as the power supplied to melt the glass, the pressure in the melting vessel and the room temperature, will be considered as disturbances. Therefore the process can be modelled as a 2-input 2-output process with disturbances.

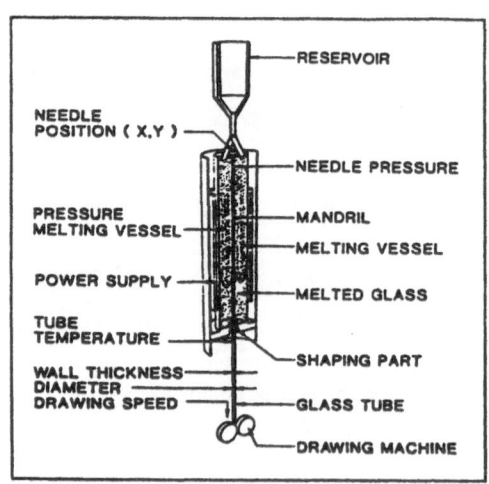

Fig. 4.3.6   The glass tube production process

Strong interactions between different inputs and outputs exist: an increase in the drawing speed causes a decrease of both the wall thickness and diameter; an increase of gas pressure will increase the diameter, but decrease the wall thickness. There are large time delays, because the tube dimensions can only be measured when the glass has considerably cooled down. Thus it is not difficult to understand why classical PID (proportional, integral and differential) control fails to work for the process.

Identification and multivariable control techniques were to be developed for this process. The objectives of the project were:

1) a reduction of the dimension variations at different working points (increasing product quality);

2) automation of change-over between different working points, and reduction of change-over time (increasing flexibility)

According to the staircase experiment, the process is reasonably linear around working points. A step change experiment shows that the impulse responses are long. This means that a FIR model needs a large number of parameters which is not convenient for control system design and implementation. So we will estimate a compact parametric model using the least-squares method.

Let us name the input/output variables as follows:

$u_1(t)$: input 1, gas pressure; $u_2(t)$: input 2, drawing speed;
$y_1(t)$: output 1, wall thickness; $y_2(t)$: output 2, diameter.

A general linear relation between the inputs and outputs can be described by a transfer function matrix

$$\begin{bmatrix} y_1(t) \\ y_2(t) \end{bmatrix} = \begin{bmatrix} G_{11}^o(q) & G_{12}^o(q) \\ G_{21}^o(q) & G_{22}^o(q) \end{bmatrix} \begin{bmatrix} u_1(t) \\ u_2(t) \end{bmatrix} + \begin{bmatrix} v_1(t) \\ v_2(t) \end{bmatrix} \qquad (4.3.5)$$

or

$$\begin{cases} y_1(t) = G_{11}^o(q)u_1(t) + G_{12}^o(q)u_2(t) + v_1(t) \\ y_2(t) = G_{21}^o(q)u_1(t) + G_{22}^o(q)u_2(t) + v_2(t) \end{cases}$$

where $G_{ij}^o(q^{-1})$ is a rational of polynomials in the delay operator q^{-1}.

First we must parametrize the model. The generalization of the difference equation model leads to a so called matrix fraction description (MFD). For a 2-input 2-output process model a left MFD is defined as

$$G(q) = \begin{bmatrix} A_{11}(q) & A_{12}(q) \\ A_{21}(q) & A_{22}(q) \end{bmatrix}^{-1} \begin{bmatrix} B_{11}(q) & B_{12}(q) \\ B_{21}(q) & B_{22}(q) \end{bmatrix} \qquad (4.3.6)$$

where $A_{ij}(q)$, $B_{ij}(q)$ are polynomials in the delay operator q^{-1}.

In general, parametrization of MIMO MFD models for identification is not an easy task which will not be discussed in this book. (In fact, this problem can be avoided in our approach; see Chapter 8.) The diagonal form MFD is, however, simple and physically appealing; then all principles and techniques for SISO process identification can be extended to the MIMO case. For the 2-input 2-output process, the diagonal form MFD is given as

$$G(q) = \begin{bmatrix} A_1(q) & 0 \\ 0 & A_2(q) \end{bmatrix}^{-1} \begin{bmatrix} B_{11}(q) & B_{12}(q) \\ B_{21}(q) & B_{22}(q) \end{bmatrix} \qquad (4.3.7)$$

Applying this description to the general model (4.3.6) yields

$$\begin{cases} A_1^\circ(q)y_1(t) = B_{11}^\circ(q)u_1(t) + B_{12}^\circ(q)u_2(t) + A_1^\circ(q)v_1(t) \\ A_2^\circ(q)y_2(t) = B_{21}^\circ(q)u_1(t) + B_{22}^\circ(q)u_2(t) + A_2^\circ(q)v_2(t) \end{cases} \qquad (4.3.8)$$

We note that in a diagonal form the model is decoupled into two two-input single-output sub-models; for each sub-model there is a common denominator polynomial. Thus the two sub-models can be estimated separately.

Let the degrees of the all the polynomials in a sub-model be equal and call this degree the order of the sub-model. Denote the orders of the two sub-models n_1 and n_2 respectively; then $[n_1, n_2]$ defines the model struc-ture. Denote the input/output data sequence as

$$Z^N := y_1(1)\ y_2(1)\ u_1(1)\ u_2(1)\ \cdots\ \cdots\ y_1(N+n)\ y_2(N+n)\ u_1(N+n)\ u_2(N+n)$$

where $n = \max\{n_1, n_2\}$. Then the least-squares estimate of the parameters of the first sub-model which minimizes the loss function

$$V_1 = \frac{1}{N} \sum_{t=n+1}^{N+n} \left(A_1(q)y_1(t) - [B_{11}(q)u_1(t) + B_{12}(q)u_2(t)] \right)^2$$

is

$$\hat{\theta}_1 = [\frac{1}{N}\Phi_1^T\Phi_1]^{-1}\frac{1}{N}\Phi_1^T\underline{y}_1 \qquad (4.3.9)$$

where

$$\underline{y}_1 = \begin{bmatrix} y_1(n+1) \\ y_1(n+2) \\ \vdots \\ y_1(n+N) \end{bmatrix}, \quad \theta_1 = \begin{bmatrix} a_{1,1} \\ \vdots \\ a_{1,n_1} \\ b_{11,1} \\ \vdots \\ b_{11,n_1} \\ b_{12,1} \\ \vdots \\ b_{12,n_1} \end{bmatrix}$$

$$\Phi_1 = \begin{bmatrix} y_1(n_1) & \cdots & y_1(1) & | & u_1(n_1) & \cdots & u_1(1) & | & u_2(n_1) & \cdots & u_2(1) \\ y_1(n_1+1) & & \vdots & | & u_1(n_1+1) & & \vdots & | & u_2(n_1+1) & & \vdots \\ \vdots & & \vdots & | & \vdots & & \vdots & | & \vdots & & \vdots \\ y_1(n_1+N) & \cdots & y_1(N+1) & | & u_1(n_1+N) & \cdots & u_1(N+1) & | & u_2(n_1+N) & \cdots & u_2(N+1) \end{bmatrix}$$

The same can be done for the second sub-model.

For the estimation of the model of this glass tube process, a PRBS experiment was performed. After the pretreatment of the data, 1269 samples are available. We shall use the first 600 samples for model estimation and the remaining 669 samples for model validation. The model structure [4, 4] is used. Fig. 4.3.7 shows the result of model validation; the relative simulation errors are 39.8% and 41.7%. We find that this model is rather inaccurate. With this poor quality we feel insecure about the applicability of the model.

Why does least-squares function well for the rolling mill process but perform poorly for the glass tube process? In the next section we are going to analyse the least-squares method and answer this question. Also, based on the analysis, ways to improve the least-squares method are highlighted.

At this point, we will congratulate you if the least-squares method does solve your problem in process modelling. You may stop reading and spend your time on other more important business. However, those readers who cannot obtain satisfactory results with the least-squares method, we encourage you to read on; this also holds for readers who would like to know more about recent progress in process identification.

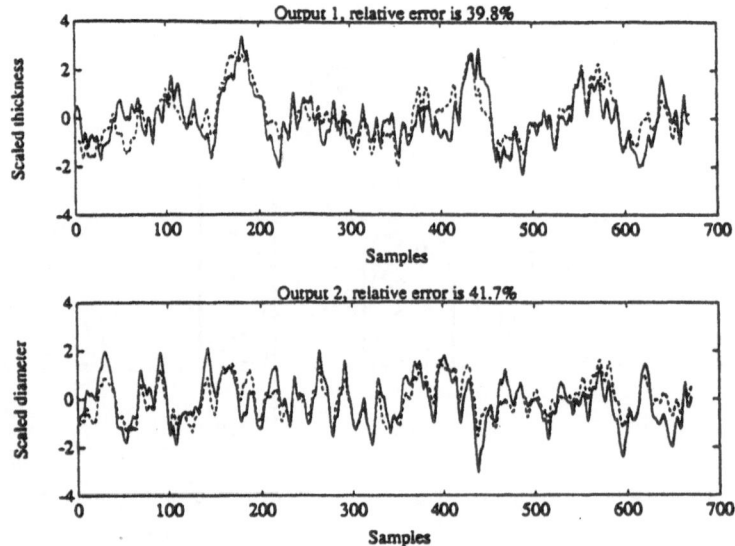

Fig. 4.3.7   Model validation for the glass tube process
"solid" := measured output; "dashed" := model output

4.4   Properties of the Least-Squares Estimator

In the previous section we have tested the least-squares method by using real life data from industrial processes. The FIR model of the rolling mill process performs very well; the transfer function model of the glass process, however, is not accurate according to our model validation. So this is a good moment to carry out some theoretical analysis on the least-squares method in order to find explanations of the behaviour of the method and to search for better methods.

Let us first look back at Section 4.1. Assume that the data are generated by the process:

$$y(t) = \varphi(t)\theta^{\circ} + e(t) \tag{4.4.1}$$

where $e(t)$ is the error, $\varphi(t)$ is the data vector

$$\varphi(t) = [x_1(t) \; x_2(t) \; \cdots \; x_n(t)]$$

and θ° is the true parameter vector

$$\theta^\circ = \begin{bmatrix} \theta^\circ_1 \\ \theta^\circ_2 \\ \vdots \\ \theta^\circ_n \end{bmatrix}$$

For the analysis let us make two other assumptions:

A1 The error term $v(t)$ is a stationary stochastic process with zero mean value $(Ev(t) = 0)$

A2 The error $e(t)$ is uncorrelated with the signals of the data vector $(E\varphi^T(t)e(t) = 0)$

Under these assumptions, the least-squares estimate $\hat{\theta}$ from (4.1.7) and the weighted least-squares estimate $\hat{\theta}_w$ from (4.1.9) are random variables. Therefore, their accuracies can be measured by a number of statistical properties such as bias, error covariance, efficiency and consistency. Common desirable properties are defined for the estimators:

Unbiased estimator: if for each sample number N

$$E\hat{\theta} = \theta^\circ$$

Consistent estimator: if

$$N \to \infty \text{ then } \hat{\theta} \to \theta^\circ \text{ with probability } 1$$

Efficient or minimum variance estimator: if for all unbiased estimators $\hat{\theta}^*$

$$\text{cov}(\hat{\theta}) \leq \text{cov}(\hat{\theta}^*) \quad \text{or} \quad \det[\text{cov}(\hat{\theta})] - \det[\text{cov}(\hat{\theta}^*)] \leq 0$$

where the covariance matrix is defined as

$$\text{cov}(\hat{\theta}) = E\{[\hat{\theta} - E(\hat{\theta})][\hat{\theta} - E(\hat{\theta})]^T\}$$

If the first and third properties hold only for $N \to \infty$, then they are called asymptotic unbiasedness and asymptotic efficiency.

Substituting (4.4.1) into (4.1.7) yields

$$\hat{\theta} = \theta^\circ + [\Phi^T\Phi]^{-1}\Phi^T\underline{e} \tag{4.4.2}$$

Taking the expectation on both sides of this equation, noting that $Ee(t) = 0$ and $Ee(t)\varphi(t-\tau) = 0$, we obtain

$$E\hat{\theta} = E\theta^\circ + E[(\Phi^T\Phi)^{-1}\Phi^T]E(\underline{e}) = \theta^\circ \tag{4.4.3}$$

Therefore *under assumption* **A1** *and* **A2** *the least-squares estimator is unbiased*. A similar proof can be given for the weighted least-squares estimator:

$$E(\hat{\theta}_w) = \theta^\circ$$

The covariance matrix of the least-squares estimator is

$$\begin{aligned}
\text{cov}(\hat{\theta}) &= E\{[\hat{\theta} - E(\hat{\theta})][\hat{\theta} - E(\hat{\theta})]^T\} \\
&= E\{[(\Phi^T\Phi)^{-1}\Phi^T\underline{e}][(\Phi^T\Phi)^{-1}\Phi^T\underline{e}]^T\} \\
&= (\Phi^T\Phi)^{-1}\Phi^T E(\underline{e}\underline{e}^T)\Phi^T(\Phi^T\Phi)^{-1} \\
&= (\Phi^T\Phi)^{-1}\Phi^T R\Phi^T(\Phi^T\Phi)^{-1}
\end{aligned} \tag{4.4.4}$$

where R is the $N \times N$ dimensional covariance matrix of the error $\underline{e}$

$$R = E\underline{e}\underline{e}^T \tag{4.4.5}$$

Similarly for the weighted least-squares estimators, we can show:

$$\text{cov}(\hat{\theta}_w) = (\Phi^T W\Phi)^{-1}\Phi^T WRW^T\Phi^T(\Phi^T W\Phi)^{-1} \tag{4.4.6}$$

If we know the error covariance R defined by (4.4.5), then it is instructive to use the weighting

$$W = R^{-1}$$

This means that we will give a data point a higher weighting when it is more reliable and a lower weighting otherwise. This choice of weighting matrix simplifies the expression of the error covariance matrix:

$$\text{cov}(\hat{\theta}_w)\Big|_{W=R^{-1}} = (\Phi^T R^{-1}\Phi)^{-1} \tag{4.4.7}$$

We shall denote the corresponding estimate as $\hat{\theta}_{mv}$

$$\hat{\theta}_{mv} = (\Phi^T R^{-1} \Phi)^{-1} \Phi^T R^{-1} \underline{y} \tag{4.4.8}$$

This estimator is called the *minimum variance* estimator (or Markov estimator) which is efficient because for any other choice of weighting matrix W

$$\text{cov}(\hat{\theta}_{mv}) \leq \text{cov}(\hat{\theta}_w)$$

We refer to Chapter 4 of Söderström and Stoica (1989), for four different proofs of this relation.

Remarks— The principle of minimum variance or efficient estimator is a good guideline for developing estimation methods. It can, however, hardly be applied in its present form to practical situations, because the error variance matrix is not available.

— When the error $e(t)$ is white noise with zero mean and variance σ^2, its covariance matrix is $\sigma^2 I$. In this case the least-squares estimator is a minimum variance estimator (efficient estimator). However, the assumption that the error is white noise is very restricted and not realistic.

Now we will show the consistency of the least-squares method. It is not difficult to verify that

$$\Phi^T \Phi = \sum_{t=1}^{N} \varphi(t)^T \varphi(t), \qquad \Phi^T \underline{y} = \sum_{t=1}^{N} \varphi(t)^T y(t)$$

Using these equations we can rewrite the least-squares estimator (4.1.7) as

$$\hat{\theta} = \left[\frac{1}{N} \sum_{t=1}^{N} \varphi(t)^T \varphi(t) \right]^{-1} \frac{1}{N} \sum_{t=1}^{N} \varphi(t)^T y(t) \tag{4.4.9}$$

Combining (4.4.1) and (4.4.9) gives an expression of the parameter error

$$\hat{\theta} - \theta^\circ = \left[\frac{1}{N} \sum_{t=1}^{N} \varphi(t)^T \varphi(t) \right]^{-1} \frac{1}{N} \sum_{t=1}^{N} \varphi(t)^T e(t) \tag{4.4.10}$$

Under weak conditions (see Chapter 2) the sums in (4.4.10) tend to the corresponding expected values as $N \to \infty$

$$\lim_{N\to\infty} \frac{1}{N} \sum_{t=1}^{N} \varphi(t)^T\varphi(t) = E\varphi(t)^T\varphi(t), \quad \lim_{N\to\infty} \frac{1}{N} \sum_{t=1}^{N} \varphi(t)^Te(t) = E\varphi(t)^Te(t) \qquad (4.4.11)$$

Hence the parameter error tends to zero, i.e., the estimator is consistent, if

$E\varphi(t)^T\varphi(t)$ is nonsingular

$E\varphi(t)^Te(t) = 0$

In most cases the first condition is satisfied and the second condition is assumption A2. Hence the consistency is proved. From (4.4.10) we understand that consistency means that the effect of disturbance can be "averaged out" by taking more measurement.

The Properties of a FIR Model

In finite impulse response (FIR) model estimation, we assume that the data is generated by

$$y(t) = \varphi(t)\theta^\circ + v(t) \qquad (4.4.12)$$

where $v(t)$ is the output disturbance which is zero mean stationary stochastic process,

$$\varphi(t) = [u(t-1) \; \cdots \; u(t-n)], \quad \theta^\circ = \begin{bmatrix} g_1^\circ \\ g_2^\circ \\ \vdots \\ g_n^\circ \end{bmatrix}$$

Assume that the length of the true impulse response is n and the input is persistently exciting of order n, then from the above analysis we can say that *the least-squares estimate of FIR model is unbiased and consistent if the disturbance v(t) is independent of the input u(t).*

Remarks— In some literature the disturbance $v(t)$ is assumed to be white noise. This assumption is too restrictive and also not necessary for the properties of unbiasedness and consistency of the FIR model.

— In practice, the independence between the input and disturbance is the consequence of open-loop identification experiments, i.e., there is no feedback from $y(t)$ to $u(t)$. In turn this implies that *when the data is collected from a closed loop experiment, the FIR estimate will be biased.*

The Bias of the Least-Squares Estimator of a Transfer Function Model

Using transfer operator models a SISO process can be described as

$$y(t) = \frac{B^\circ(q)}{A^\circ(q)}u(t) + v(t) \qquad (4.4.13)$$

where $A^\circ(q)$, $B^\circ(q)$ are the true denominator and numerator polynomials of the transfer operator, $v(t)$ is the output disturbance which is a stationary stochastical process with zero mean. Assume that the data are collected from an open loop experiment so that the the disturbance $v(t)$ and input $u(t)$ are independent.

In the least-squares method one rewrites this relation in the equation error (linear regression) model

$$y(t) = \varphi(t)\theta^\circ + e(t) \qquad (4.4.14)$$

where

$$\varphi(t) = [-y(t-1) \cdots -y(t-n) \ u(t-1) \cdots u(t-n)]$$

$$\theta^\circ = \begin{bmatrix} a^\circ_1 \\ \vdots \\ a^\circ_n \\ b^\circ_1 \\ \vdots \\ b^\circ_n \end{bmatrix}$$

We note that, unfortunately, the data vector $\varphi(t)$ is now correlated with the equation error $e(t)$, because the output $y(t)$ contains the disturbance $v(t)$ and $e(t)$ is correlated with $v(t)$ via

$$e(t) = A^\circ(q)v(t)$$

This implies that condition A2 no longer holds for a transfer function model; therefore, *in general the least-squares estimator of transfer function model is biased and not consistent.*

One may ask under what conditions will the least-squares estimator of transfer function model be unbiased and consistent. To answer this question let us do some more analysis. The least-squares estimator is given by

$$\hat{\theta} = \left[\frac{1}{N}\sum_{t=n+1}^{N+n}\varphi(t)^T\varphi(t)\right]^{-1}\frac{1}{N}\sum_{t=n+1}^{N+n}\varphi(t)^Ty(t) \qquad (4.4.15)$$

Combining (4.4.14) and (4.4.15) we obtain

$$\hat{\theta} - \theta^{\circ} = \left[\frac{1}{N}\sum_{t=n+1}^{N+n}\varphi(t)^T\varphi(t)\right]^{-1}\frac{1}{N}\sum_{t=n+1}^{N+n}\varphi^T(t)e(t)$$

$$= \left[\frac{1}{N}\sum_{t=n+1}^{N+n}\varphi(t)^T\varphi(t)\right]^{-1}\frac{1}{N}\sum_{t=n+1}^{N+n}\begin{bmatrix} -y^{\circ}(t-1) \\ \vdots \\ -y^{\circ}(t-n) \\ u(t-1) \\ \vdots \\ u(t-n) \end{bmatrix}e(t) + \left[\frac{1}{N}\sum_{t=n+1}^{N+n}\varphi(t)^T\varphi(t)\right]^{-1}\frac{1}{N}\sum_{t=n+1}^{N+n}\begin{bmatrix} v(t-1) \\ \vdots \\ v(t-n) \\ 0 \\ \vdots \\ 0 \end{bmatrix}e(t)$$

where $y^{\circ}(t)$ is the noise-free output. When $N \to \infty$, the first term of the right hand side will tend to zero due to open-loop experiment; the second term will tend to

$$\left[E\varphi(t)^T\varphi(t)\right]^{-1}\begin{bmatrix} Ev(t-1)e(t) \\ \vdots \\ Ev(t-n)e(t) \\ 0 \\ \vdots \\ 0 \end{bmatrix} \qquad (4.4.16)$$

This is the asymptotic bias of the least-squares estimator, which will be zero only if the equation error $e(t)$ is white noise. This is an *extremely* restrictive condition and it cannot be satisfied in practical situations (think of the two industrial processes described in the previous section). Even if the output disturbance is white noise, which might happen if the disturbance consists of measurement noise only, the equation error will still be a coloured noise because $e(t)$ is filtered $v(t)$ ($e(t) = A^{\circ}(q)v(t)$).

Assuming a white equation error is not realistic; one must be extremely lucky to encounter such a situation. However, trying to obtain white noise residuals is a way to an unbiased and consistent estimator (see Chapter 5).

Error Distribution in the Frequency Domain

So far we have analysed the properties of the least-squares method in the parameter space. A control engineer, however, is more concerned about the model behaviour in the frequency domain, because for model applications such as control and simulation, the parameter set of a discrete-time model is merely a vehicle to arrive at a good transfer function estimate. If the model order is lower than the true process order, which often happens in applications, it does not make sense to talk about parameter errors; on the other hand, the error of the transfer function is still well defined.

Assume that an open-loop experiment has been performed. Let us write the loss function of the least-squares estimate for transfer function models as

$$V_{LS} = \frac{1}{N} \sum_{t=n+1}^{N+n} \varepsilon(t)^2 = \frac{1}{N} \sum_{t=n+1}^{N+n} \left(A(q)y(t) - B(q)u(t) \right)^2$$

When $N \to \infty$

$$V_{LS} \to E \left(A(q)y(t) - B(q)u(t) \right)^2$$

$$[\text{apply } (4.4.13)] = E \left(A(q)[G^o(q)u(t) + v(t)] - B(q)u(t) \right)^2$$

$$= EA^2(q) \left\{ \left(G^o(q) - \frac{B(q)}{A(q)} \right) u(t) + v(t) \right\}^2$$

$$[Eu(t)v(t) = 0] = EA^2(q) \left(G^o(q) - \frac{B(q)}{A(q)} \right)^2 u^2(t) + EA^2(q)v^2(t)$$

Applying Parseval's formula (see Chapter 2) we obtain the asymptotic least-squares loss function in the frequency domain

$$V_{LS} \to \frac{1}{2\pi} \int_{-\pi}^{\pi} \left(G^o(e^{i\omega}) - \frac{B(e^{i\omega})}{A(e^{i\omega})} \right)^2 A^2(e^{i\omega}) \Phi_u(\omega) d\omega + \frac{1}{2\pi} \int_{-\pi}^{\pi} A^2(e^{i\omega}) \Phi_v(\omega) d\omega \quad (4.4.17)$$

where $G^o(e^{i\omega})$ is the true transfer function, $\Phi_u(\omega)$ and $\Phi_v(\omega)$ are the spectra of the input and the disturbance. The first term of (4.4.17) shows that in the least-squares loss function the filter $A(q)^2$ is part of the weighting in the transfer function approximation. For most of the industrial processes, $A(q)$ is a highpass filter because $1/A(q)$ is lowpass. Therefore, *the effect of $A(q)^2$ is to impose heavy weighting at high frequencies.* This is certainly not desirable in model applications such as control and simulation.

Now we are in a position to give theoretical explanations to what happened in the two case studies:

1) The model of the rolling mill is accurate because the least-squares estimate of FIR model is unbiased and consistent.

2) The quality of the glass tube process is poor because the least-squares estimate of transfer function model is biased; and the criterion emphasizes the fit of the transfer function at high frequencies at the cost of the fit at middle and low frequencies.

Logically, ways to improve the estimate of transfer function models are to remove the bias and to change the frequency weighting. For these purposes different methods, such as generalized least-squares method, instrumental variable method, output error method and prediction error method, have been proposed and studied in the past two decades. An overview of these methods will be given in the next chapter.

4.5 Conclusions

In this chapter we have introduced the least-squares principle and we have shown how to use it in linear process identification. The method has been tested by two industrial processes, where it succeeded in the first application but failed in the second one. Then theoretical explanations were given for the two very different performances of the method. To end this chapter the following conclusions can be drawn:

• The least-squares method is simple to understand and easy to use due to the existence of the closed solution.

• The least-squares method will deliver an unbiased and consistent estimate if the process can be properly described by a FIR model, and if the data are from an open-loop experiment.

• The least-squares method can deliver a good transfer function model if the level of the disturbance is low and the model order of the process is correct. Otherwise, the least-squares estimator is biased and the model fit at high frequencies is over-emphasized.

When the level of the disturbance increases, or if the used model order is lower than the true order of the process, or when the experiment is carried out in a closed loop, the performance of the least-squares method can be poor; other methods need to be used to obtain a good process model.

CHAPTER 5

EXTENSIONS OF THE LEAST-SQUARES METHOD

In the previous chapter we have studied the least-squares (LS) method. From the two case studies and the analysis we have learnt that the LS estimates of the parameters of a transfer operator or difference equation model are biased (not consistent); and the model fit in the frequency domain is not always good, for the high frequencies are over-emphasized. In the last two decades, much research has been done to modify and to extend the LS method in order to arrive at a consistent estimator. In this chapter, we will discuss various ways of modifying the LS method.

We shall start from the frequency domain considerations. In this way, we can arrive at the so called Steiglitz-McBride method (Section 5.1). In Section 5.2 the output error method is studied. The instrumental variable (IV) method is introduced in Section 5.3. These methods can work well for an open loop experiment but will yield biased estimates when closed loop data are used. This problem can be overcome by a prediction error method which is studied in Section 5.4. In concluding this chapter we shall point out some practical limitations of these methods, especially for control applications.

5.1 Modifying the Frequency Weighting by Prefiltering

In control system design, especially in digital controller design, we are more concerned about the errors of the transfer function in the frequency domain than the errors of parameters. Let us write down again the least-squares loss function in the frequency domain as $N \to \infty$ (4.4.17)

$$V_{LS} = \frac{1}{2\pi} \int_{-\pi}^{\pi} (G^\circ(e^{i\omega}) - \frac{B(e^{i\omega})}{A(e^{i\omega})})^2 A^2(e^{i\omega}) \Phi_u(\omega) d\omega + \frac{1}{2\pi} \int_{-\pi}^{\pi} A^2(e^{i\omega}) \Phi_v(\omega) d\omega \quad (5.1.1)$$

We see that the transfer function error is weighted by the frequency weighting $[A^2(e^{i\omega}) \Phi_u(\omega)]$.

Suppose that the desired frequency weighting is given as $W^2(e^{i\omega})$. Then based on the expression (5.1.1) we have at least two ways to realize this: *input design* and *prefiltering*.

When using the idea of input design, one should choose a test input such that its spectrum $\Phi_u(\omega)$ has the form

$$\Phi_u(\omega) = W^2(e^{i\omega})/A^2(e^{i\omega})$$

However, this cannot be done in a precise way because $A(q)$ is not known before parameter estimation. One may ask why not perform identification experiment and parameter estimation repeatedly or iteratively. The answer is that doing an identification experiment is often a very costly business. Anyway, from this discussion we have learnt that the input design is related to the identification method.

If the input spectrum is the desired frequency weighting

$$\Phi_u(\omega) = W^2(e^{i\omega})$$

then we can eliminate the effect of $A^2(e^{i\omega})$ in the least-squares method by prefiltering. Denote $L(q)$ as a stable filter and perform prefiltering on the input/output data:

$$\begin{cases} u_f(t) = L(q)u(t) \\ y_f(t) = L(q)y(t) \end{cases}$$

Then applying the least-squares method on the filtered data we have

$$\hat{\theta} = [\Phi_f^T \Phi_f]^{-1} \Phi_f^T y_f \qquad (5.1.2)$$

where

$$\theta = \begin{bmatrix} a_1 \\ \vdots \\ a_n \\ b_1 \\ \vdots \\ b_n \end{bmatrix}, \quad y_f = \begin{bmatrix} y_f(n+1) \\ y_f(n+2) \\ \vdots \\ y_f(n+N) \end{bmatrix}, \quad \Phi_f = \begin{bmatrix} -y_f(n) & \cdots -y_f(1) & | & u_f(n) & \cdots u_f(1) \\ -y_f(n+1) & \vdots & | & u_f(n+1) & \vdots \\ \vdots & \vdots & | & \vdots & \vdots \\ -y_f(n+N) \cdots -y_f(N+1)| & u_f(n+N) \cdots u_f(N+1) \end{bmatrix}$$

which minimizes the following loss function

$$V_{FLS} = \frac{1}{N} \sum_{t=n+1}^{N+n} \varepsilon_f(t)^2 = \frac{1}{N} \sum_{t=n+1}^{N+n} [A(q)y_f(t) - B(q)u_f(t)]^2 \qquad (5.1.3)$$

Assuming that the true process is

$$y(t) = \frac{B^o(q)}{A^o(q)}u(t) + v(t) \qquad (5.1.4)$$

and following the same derivation as for (5.1.1) (see Section 4.4), we can show that as $N \to \infty$

$$V_{FLS} = \frac{1}{2\pi} \int_{-\pi}^{\pi} (G^{\circ}(e^{i\omega}) - \frac{B(e^{i\omega})}{A(e^{i\omega})})^2 L^2(e^{i\omega}) A^2(e^{i\omega}) \Phi_u(\omega) d\omega$$

$$+ \frac{1}{2\pi} \int_{-\pi}^{\pi} L^2(e^{i\omega}) A^2(e^{i\omega}) \Phi_v(\omega) d\omega \qquad (5.1.5)$$

In order to cancel $A^2(e^{i\omega})$ in the weighting the filter should be

$$L(q) = 1/A(q)$$

Of course, $A(q)$ is unknown. But this can be solved by the following iteration procedure:

1) To start, perform a normal least-squares estimation, and denote the estimate as $\hat{G}^1(q) = \hat{B}^1(q)/\hat{A}^1(q)$.

2) Denote $\hat{G}^{(k)}(q) = \hat{B}^k(q)/\hat{A}^k(q)$ as the estimate of model at iteration k. First filter the input $u(t)$ and the output $y(t)$ by $1/\hat{A}^k(q)$. Then apply the least-squares method on the filtered data.

If the iteration converges, we will get an estimate which is a local minimum of the following loss function

$$V_{\infty} = \frac{1}{2\pi} \int_{-\pi}^{\pi} (G^{\circ}(e^{i\omega}) - \frac{B(e^{i\omega})}{A(e^{i\omega})})^2 \Phi_u(\omega) d\omega + \frac{1}{2\pi} \int_{-\pi}^{\pi} \Phi_v(\omega) d\omega \qquad (5.1.6)$$

Note that this loss function holds also when the model order is lower than the process order. This algorithm was proposed by Steiglitz and McBride (1965), where their motivation was to find a simple algorithm for minimizing the output error loss function (5.2.2). A block diagram of their method is shown in Fig. 5.1.1. The method is numerically simple and has a straight-forward interpretation liked by engineers. In our experience the iteration converges very well.

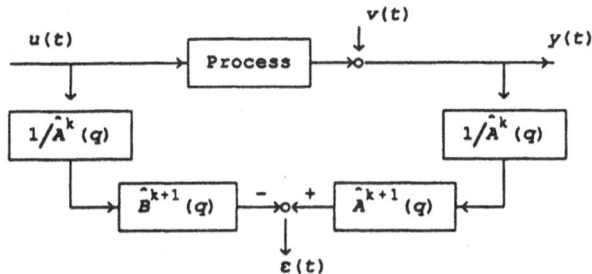

Fig. 5.1.1   Error generation of Steiglitz-McBride method

5.2   A Natural Choice of Criterion—Output Error Methods

Let us now go back to a fundamental problem — the choice of criterion (loss function) for model approximation. We have mentioned before that equation error criterion used in the least-squares method is *not* a natural choice; we choose it because it is easy to use and simple to comprehend. We have proposed the output error criterion for model validation because it is closer to model applications as required in control and simulation. Now we will use this criterion for model estimation.

Denote the process model as

$$G(q) = \frac{B(q)}{A(q)} = \frac{b_1 q^{-1} + \cdots b_n q^{-n}}{1 + a_1 q^{-1} + \cdots a_n q^{-n}}$$

The output error is defined as

$$\varepsilon_{oe}(t) = y(t) - \frac{B(q)}{A(q)}u(t) = y(t) - \hat{y}(t) \tag{5.2.1}$$

Given the input/output data sequence as

$$Z^N := y(1) \ u(1) \ \cdots \ \cdots \ y(N) \ u(N)$$

and assuming that the process order n is known. Then the output error method estimates parameters by minimizing the loss function

$$V_{OE}^N = \frac{1}{N} \sum_{t=1}^{N} \varepsilon_{oe}(t)^2 \tag{5.2.2}$$

The block diagram for generating the output error is given in Fig. 5.2.1.

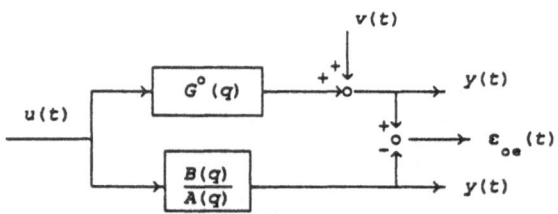

Fig. 5.2.1 Output error generation

We note that the output error $\varepsilon_{oe}(t)$ is *nonlinear* in the parameters of the $A(q)$ polynomial; the consequence of this nonlinearity is that there exists no analytical solution to this minimization problem. Therefore, a numerical search algorithm is needed to find a minimum; this is much more time consuming than the least-squares solution. Problems such as local minima and non-convergence can occur. Also the theoretical analysis becomes more involved.

Before showing how to calculate the estimate, let us study the properties of the output error method for open loop data. First the consistency is shown. Assume that the true process is given by (5.1.4) where $v(t)$ is stationary with zero mean, that the model order n is correct, that the input is persistently exciting with order $2n$, and that the minimization of (5.2.2) converges to the global minimum for all N. Denote the model error as

$$\Delta G(q) = G^\circ(q) - \hat{G}(q)$$

then if $N \to \infty$

$$V_{OE}^N \to EV_{OE} = E[\Delta G(q)u(t) + v(t)]^2$$
$$= \Delta G(q)^2 Eu(t)^2 + 2\Delta G(q)E[u(t)v(t)] + Ev(t)^2$$
$$= \Delta G(q)^2 Eu(t)^2 + Ev(t)^2 \qquad (5.2.3)$$

The last equality in (5.2.3) is due to an open loop experiment. In such case the test input is not correlated with the output disturbance. If the model order is correct and the minimization finds the global minimum, we have

$$\Delta G(q)^2 \to 0 \text{ as } N \to \infty$$

which implies that

$$\hat{G}(q) \to G^o(q) \text{ as } N \to \infty \qquad (5.2.4)$$

This is equivalent to saying that

$$\hat{\theta} \to \theta^o(q) \text{ as } N \to \infty \qquad (5.2.5)$$

where

$$\theta = \begin{bmatrix} a_1 \\ \vdots \\ a_n \\ b_1 \\ \vdots \\ b_n \end{bmatrix}$$

Hence the output error method is consistent under relatively weak conditions. In plain words, when using the output error method the effect of the output disturbance can be averaged out, provided that the input is not correlated with the disturbance (in the case of open loop experiment).

If the model order is lower than the true one (undermodelling), which can often happen in applications, the model will be biased; we shall see how the model error is distributed for the output error method model. Applying Parseval's formula on (5.2.3) we obtain for $N \to \infty$,

$$V_{OE}^N \to \frac{1}{2\pi} \int_{-\pi}^{\pi} \left(G^o(e^{i\omega}) - \frac{B(e^{i\omega})}{A(e^{i\omega})} \right)^2 \Phi_u(\omega)d\omega + \frac{1}{2\pi} \int_{-\pi}^{\pi} \Phi_v(\omega)d\omega \qquad (5.2.6)$$

Hence using the output error criterion, the error of transfer function is weighted by the input spectrum, which can be manipulated by input design.

Summarizing, in a open-loop operation, the output error method is consistent if the model order is correct. If the model order is lower than the true one, the error of the transfer function estimate is weighted by the input spectrum. In other words, if the order is correct, the identified model will be accurate; if the model order is too low, the model error can be effectively affected by input design. Remember, however, that a precondition for these nice properties is the global convergence of the numerical search algorithm used. In practical situations this is not always guaranteed.

A simple minimization algorithm

The loss function V_{OE} is nonlinear in the parameters of $A(q)$ and it must be minimized using a numerical search routine. There exists a large variety of numerical methods for optimization like steepest descent, Newton-Raphson, Marquardt, and stochastic approximation, to name a few. We will not, in this book, become too involved in numerical optimization problems. There are two reasons for this. First, many methods are available as standard routines from different computer packages for identification and control design. Secondly, numerical difficulties can be reduced to a great extend in our identification methods (see Chapters 6, 7 and 8). Here we will introduce the Gauss-Newton method which is well suited for the sum of squares criterion. The purpose is to give some insight into numerical optimization.

Denote $\hat{\theta}^k$ as the estimate at the kth iteration. Assume that $\hat{\theta}^k$ is close to the minimum. For parameter vectors close to $\hat{\theta}^k$ we can approximate the output error residual by truncating the higher order terms of its Taylor expansion so that

$$\varepsilon_{oe}(t,\theta) \approx \varepsilon(t,\hat{\theta}^k) + \left.\frac{\partial \varepsilon(t,\theta)}{\partial \theta^T}\right|_{\theta=\hat{\theta}^k}(\theta - \hat{\theta}^k)$$

$$= \varepsilon(t,\hat{\theta}^k) + \varphi^k(t)(\theta - \hat{\theta}^k)$$

$$= -[\varphi^k(t)\hat{\theta}^k - \varepsilon(t,\hat{\theta}^k)] + \varphi^k(t)\theta \qquad (5.2.7)$$

Here $\varphi^k(t)$ is the gradient of the output error residual (5.2.1)

$$\varphi^k(t) = \left.\frac{\partial \varepsilon(t,\theta)}{\partial \theta^T}\right|_{\theta=\hat{\theta}^k}$$

$$= \left[\left.\frac{\partial \varepsilon(t,\theta)}{\partial a_1}\right|_{a_1=\hat{a}_1} \cdots \left.\frac{\partial \varepsilon(t,\theta)}{\partial a_n}\right|_{a_n=\hat{a}_n} \left.\frac{\partial \varepsilon(t,\theta)}{\partial b_1}\right|_{b_1=\hat{b}_1} \cdots \left.\frac{\partial \varepsilon(t,\theta)}{\partial b_n}\right|_{b_n=\hat{b}_n}\right]$$

$$= \left[\frac{\hat{B}^k(q)}{\hat{A}^k(q)^2}u(t-1) \cdots \frac{\hat{B}^k(q)}{\hat{A}^k(q)^2}u(t-n) \frac{-1}{\hat{A}^k(q)}u(t-1) \cdots \frac{-1}{\hat{A}^k(q)}u(t-n)\right]$$

Under this approximation, we find that the error is linear in the parameters; hence the least-squares method can be used to find the refined estimate. Comparing (5.2.7) to linear regression equation (4.2.6), we find that now $[\varphi^k(t)\hat{\theta}^k - \varepsilon(t,\hat{\theta}^k)]$ takes the position of $y(t)$ and $\varphi^k(t)$ takes the position of $\varphi(t)$. Therefore the new estimate is given by

$$\hat{\theta}^{k+1} = \left[\sum_{t=1}^{N} [\varphi^k(t)]^T \varphi^k(t) \right]^{-1} \sum_{t=1}^{N} [\varphi^k(t)]^T [\varphi^k(t)\hat{\theta}^k - \varepsilon(t,\hat{\theta}^k)]$$

$$= \hat{\theta}^k - \left[\sum_{t=1}^{N} [\varphi^k(t)]^T \varphi^k(t) \right]^{-1} \sum_{t=1}^{N} [\varphi^k(t)]^T \varepsilon(t,\hat{\theta}^k) \qquad (5.2.8)$$

Some conditions should be satisfied in order to make the algorithm work. First the models from each iteration must be stable, because otherwise the elements of $\varphi^k(t)$ will be unbounded. If an unstable intermediate model is obtained, we can approximate it by a stable one. Secondly $\hat{A}^k(q)$ and $\hat{B}^k(q)$ must have no common factor (coprime) and the input must be persistently exciting with order $2n$. This is to ensure that the matrix $\left[\sum_{t=1}^{N} [\varphi^k(t)]^T \varphi^k(t) \right]$ is nonsingular. Obviously, these conditions are not very restrictive for practical applications.

The least-squares estimate, or the estimate from the Steiglitz-McBride method, can be used to start the iteration.

This method is also called a quasilinearization algorithm due to the approximation in (5.2.7).

Remark—There is a slight change in data arrangement: $N+n$ is replaced by N and the summation is over 1 to N. This is for the ease of notation and will be adopted in the rest of the book. If $N \gg n$, the effect of this arrangement is negligible. When implementing the estimation algorithm, the old arrangement can be used, because it is more accurate.

The output error method versus the Steiglitz-McBride method

Some readers might have found similarities between the output error method and the Steiglitz-McBride method of the previous section; some readers might even think that they are identical. This is not surprising because the Steiglitz-McBride method was proposed in order to approximate the output error method. There exists, however, a conceptual difference between the two methods. The output error method aims at minimizing the output error loss function (5.2.2); the Steiglitz-McBride method is an *ad hoc* iteration scheme which is not a minimization procedure, in other words, the output loss function will not necessarily decrease at each iteration. Therefore, if the latter converges we in general, will find a local minimum of the output error loss function. Söderström and Stoica (1981) have shown that if the disturbance $v(t)$ is white noise and if the model order is correct, the Steiglitz-McBride iteration will converge to the global minimum of the loss function (5.2.2). The requirement for a white noise disturbance is a very restrictive condition. Söderström and Stoica (1982) have shown that if the test input $u(t)$ (not disturbance) is white noise, then output error method will converge to the global minimum of the loss function; otherwise, local minima may exist. In general, under the same conditions, an output error model of a process will be more accurate than a Steiglitz-McBride model.

The Steiglitz-McBride method is simple to implement and readily understood. From our practical experience, we have yet to see an example where the difference between the output error model and the Steiglitz-McBride model is significant.

Example 5.2.1   Model validation and order selection

In Section 4.2.3 we have suggested that simulation be used as a tool for model order selection and model validation. Now that the output error criterion is the same as the criterion of model validation, we should expect a result different from that of the least-squares method.

The same data generated in Example 4.2.1 are used here, the output error method is used to estimate models of increasing orders. The first 500 samples are used for estimation; the second 500 samples are used for model validation and order selection. The variations of the loss function V_{oe} of the output error models as a function of model order for different noise-to-signal ratios are plotted in Fig. 5.2.1. For comparison, the loss function V_{OE} of the least-squares models are also plotted in the same plots.

We find that the true order can be found in all cases. Also when first order is used (undermodelling), the output error models will give a better approximation than the least-squares models. The problem of local minimum does not occur in this example, because the input is white noise. If we look at the plots carefully, we see that, in the case of $N/S = 100\%$, the loss function increases with the order. This is due to the fact that the variance of the model (caused by the disturbance) increases with model order; see Chapter 7 for explanations.

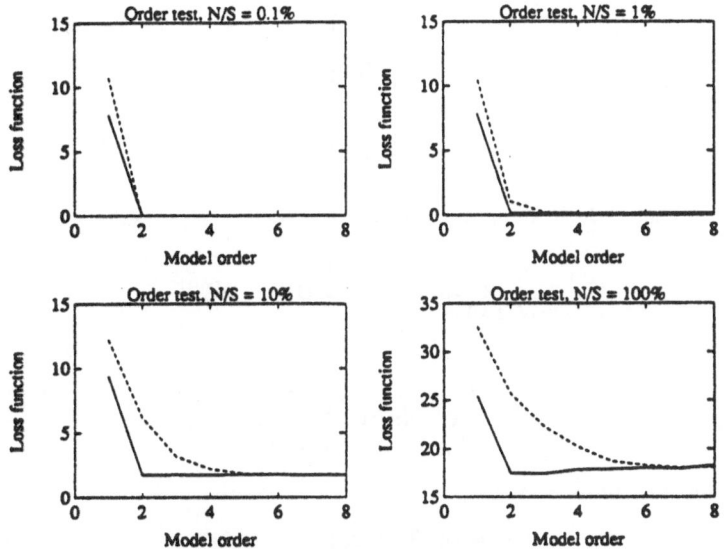

Fig. 5.2.1 Order test for different noise-to-signal ratios.
"solid line" := the loss functions of output error models;
"dashed line" := the loss functions of least-squares models.

5.3 Using Correlation Techniques—Instrumental Variable (IV) Methods

Another way of modifying the least-squares method to overcome the bias problem is to use *instrumental variable* (IV) methods. We shall motivate this group of methods by revisiting the LS method. Write down the equation error model used in LS method:

$$A(q)y(t) = B(q)u(t) + \varepsilon(t) \tag{5.3.1}$$

We have shown that this model can be written as the linear regression form

$$y(t) = \varphi(t)\theta + \varepsilon(t) \tag{5.3.2}$$

where $\varphi(t)$ is the data vector

$$\varphi(t) = [-y(t-1), \cdots -y(t-n) \quad u(t-1) \cdots u(t-n)]$$

and θ is the parameter vector

$$\theta = \begin{bmatrix} a_1 \\ \vdots \\ a_n \\ b_1 \\ \vdots \\ b_n \end{bmatrix}$$

Then the LS estimate which minimizes the loss function

$$V_{LS} = \frac{1}{N} \sum_{t=1}^{N} \varepsilon(t)^2 = \frac{1}{N} \sum_{t=1}^{N} [y(t) - \varphi(t)\theta]^2$$

is

$$\hat{\theta} = \left[\frac{1}{N} \sum_{t=1}^{N} \varphi^T(t)\varphi(t)\right]^{-1} \frac{1}{N} \sum_{t=1}^{N} \varphi^T(t)y(t)$$

Assume that the true process is given by

$$y(t) = \varphi(t)\theta° + e(t) \tag{5.3.3}$$

cf. (4.4.13) and (4.4.14). The parameter errors can be determined by

$$\hat{\theta} - \theta° = \left[\frac{1}{N} \sum_{t=1}^{N} \varphi^T(t)\varphi(t)\right]^{-1} \frac{1}{N} \sum_{t=1}^{N} \varphi^T(t)e(t)$$

When N tends to infinity this becomes

$$\hat{\theta} - \theta° = [E\varphi^T(t)\varphi(t)]^{-1} E\varphi^T(t)e(t)$$

This is the asymptotic bias which will not be zero unless $e(t)$ is white noise.

Now assume that $Z(t)$ is a signal vector with dimension $1 \times 2n$ (the same dimension as $\varphi(t)$), the entries of which are signals correlated with the input $u(t)$ but *uncorrelated* with the equation disturbance $e(t)$. Then we may try to estimate the parameters by exploiting this property. Correlating the both sides of equation (5.3.2) with $Z(t)^T$ using the data for $t = 1, 2, \cdots$ $N+n$ we obtain

$$\frac{1}{N} \sum_{t=1}^{N} Z(t)^T y(t) = \frac{1}{N} \sum_{t=1}^{N} Z(t)^T \varphi(t)\theta° + \frac{1}{N} \sum_{t=1}^{N} Z(t)^T \varepsilon(t) \tag{5.3.4}$$

Since $Z(t)$ is assumed uncorrelated with equation disturbance $e(t)$, for large N we have:

$$\frac{1}{N} \sum_{t=1}^{N} Z(t)^T \epsilon(t) \approx 0$$

Then equation (5.3.4) gives rise to an estimate

$$\hat{\theta} = \left[\frac{1}{N} \sum_{t=1}^{N} Z(t)^T \varphi(t) \right]^{-1} \frac{1}{N} \sum_{t=1}^{N} Z(t)^T y(t) \tag{5.3.5}$$

This is the so called *basic instrumental variable* (IV) estimate of θ. The elements of the vector $Z(t)$ are called the *instrument variables*. If $Z(t)$ is replaced by $\varphi(t)$ we can rederive the LS estimate.

It is easy to verify that the parameter errors are given by

$$\hat{\theta} - \theta^\circ = \left[\frac{1}{N} \sum_{t=1}^{N} Z(t)^T \varphi(t) \right]^{-1} \frac{1}{N} \sum_{t=1}^{N} Z(t)^T e(t)$$

Due to the definition of $Z(t)$, the errors tend to zero when N tends to infinity, and the instrumental variable estimator is therefore consistent.

There are many ways to choose the instrument variable vector $Z(t)$. All the choices will give consistent estimates; the differences lie in the corresponding covariances. One possibility for choosing the instruments is the following:

$$Z(t) = [-\alpha(t-1) \cdots -\alpha(t-n) \; u(t-1) \cdots u(t-n)] \tag{5.3.6}$$

where the signal $\alpha(t)$ is obtained by filtering the input

$$\alpha(t) = \frac{D(q)}{C(q)} u(t) \tag{5.3.7}$$

Again, the polynomials $C(q)$ and $D(q)$ can be chosen in many ways. One special choice is to let $C(q)$ and $D(q)$ be the LS estimates of $A(q)$ and $B(q)$ respectively. Another special choice is to use delayed inputs

$$Z(t) = [u(t-1) \cdots u(t-1) \cdots u(t-2n)] \tag{5.3.8}$$

The block diagram of IV-methods is shown in Fig. 5.3.1

80

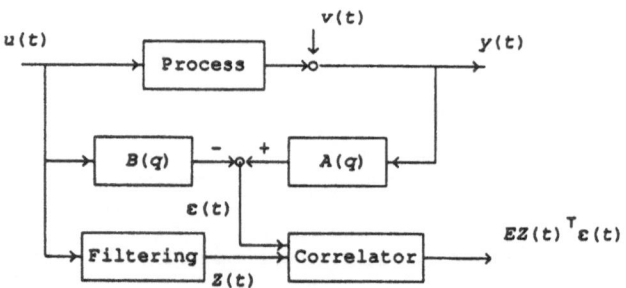

Fig. 5.3.1   Instrumental variable methods

Logically, one will ask what the optimal choice of instrumental variables is. It can be shown (Söderström and Stoica, 1989, Chapter 8) that the optimal $Z(t)$ which gives the smallest covariance of $\hat{\theta}$ can be generated by the true transfer function $G^{o}(q)$ and the true disturbance filter $H^{o}(q)$. Because these true models are unknown, some iterations are inevitable for approximating an optimal $Z(t)$. We say *an* optimal $Z(t)$ instead of *the* optimal $Z(t)$ because a nonsingular linear transformation of $Z(t)$ has no influence on the estimate. It can easily be seen from (5.3.5) that a change of instruments from $Z(t)$ to $TZ(t)$, where T is a nonsingular matrix, will not change $\hat{\theta}$.

Model validation and order selection for IV-method can be done by simulation as studied before.

The advantages of the IV-methods are: numerical simplicity; and the estimate is consistent. So in general this method is superior to the least-squares method. However, for the implementation of an optimal IV, iterations are needed; this means that the advantage of numerical simplicity will be lost. For the higher model accuracy, one may use a prediction error method which will be studied in the following section.

If the model order is lower than the process order, it is not clear in what sense an IV estimate approximates the process transfer function. In this case, we would prefer to use an output error method, because we can easily influence the model misfit in the frequency domain.

So far we have assumed that experiments are carried out where the process to be identified is in open loop. In a closed loop experiment, all the above three methods will be biased, hence not consistent, because the process input $u(t)$ is *correlated* with the disturbance through feedback. It is possible, however, to modify the IV method by using an IV vector $Z(t)$ in which the process input is replaced by an external input (input from outside the loop) which is not correlated with the disturbance; see Söderström and Stoica (1989).

5.4 Obtaining White Residuals—Prediction Error Methods

In the development of the Steiglitz-McBride method, the output error method and the instrumental variable methods, we have aimed at obtaining a good estimate of the process model. Under realistic conditions, these methods can give consistent estimates of the process transfer function, which is a big step forward from the biased least-squares method. However, if we want our model to do a good job in control applications, we have to solve at least two more problems: (1) to maintain the consistency in the case of closed loop experiments and (2) to estimate a disturbance model. The first is needed again for a good process model; and the second is for optimal disturbance reduction. Prediction error methods can meet these requirements. Moreover, with the three methods described above we have not yet obtained a minimum variance or efficient estimate. For this we should study prediction error methods, for they can give asymptotically efficient estimates.

Viewing the Least-Squares Method as a Prediction Error Method

Rewrite the equation error (linear regression) model (5.3.2) as

$$y(t) = -a_1 y(t-1) - \cdots - a_n y(t-n) + b_1 u(t-1) + \cdots + b_n u(t-n) + \varepsilon(t)$$
$$= \varphi(t)\theta + \varepsilon(t)$$
$$= [1 - A(q)]y(t) + B(q)u(t) + \varepsilon(t) \tag{5.4.1}$$

Neglecting the equation error $\varepsilon(t)$ in this model, one can predict the output at time t using *both* the input and the output at t-1, t-2, $\cdots$ as

$$\hat{y}(t) = \varphi(t)\theta$$
$$= [1 - A(q)]y(t) + B(q)u(t) \tag{5.4.2}$$

Hence the equation error

$$\varepsilon(t) = y(t) - y(t) \tag{5.4.3}$$

can be interpreted as a prediction error. See Fig. 5.4.1 for the block diagram of this interpretation. Therefore, the LS method determines the parameters such that the sum of squared prediction errors is minimized. We know from Chapter 4 that the LS estimate will be unbiased and efficient (minimum variance) if the equation disturbance $e(t)$ in (5.3.3) is zero mean white noise with constant variance R. In practice, however, the equation error disturbance $e(t)$ is not white noise, which causes the bias of the LS estimate.

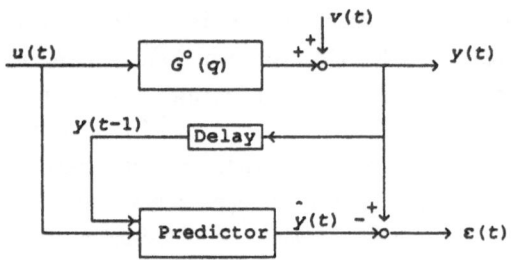

Fig. 5.4.1. Viewing the LS method as a prediction error method

The idea behind different prediction error methods is to model the equation (or output) disturbance using shaping filters in order to arrive at a consistent and efficient estimate. Different forms of disturbance filters leads to different names for methods in the identification literature.

5.4.1 Generalized Least-Squares (GLS) Method

This method was proposed by Clarke (1967), where he extended the equation error model and assumed that the true process is given by

$$A^\circ(q)y(t) = B^\circ(q)u(t) + \frac{1}{D^\circ(q)}\xi(t) \qquad (5.4.4)$$

or

$$y(t) = \frac{B^\circ(q)}{A^\circ(q)}u(t) + \frac{1}{A^\circ(q)D^\circ(q)}\xi(t)$$

where

$$\begin{cases} A^\circ(q) = 1 + a_1^\circ q^{-1} + \cdots + a_n^\circ q^{-n} \\ B^\circ(q) = \qquad b_1^\circ q^{-1} + \cdots + b_n^\circ q^{-n} \end{cases}, \quad D^\circ(q) = 1 + d_1^\circ q^{-1} + \cdots + d_{n_d}^\circ q^{-n_d}$$

and $\xi(t)$ is white noise with zero mean and variance R. So the equation disturbance is assumed to be an AR (autoregressive) process. Rewrite equation (5.4.4) as

$$D^\circ(q)A^\circ(q)y(t) = D^\circ(q)B^\circ(q)u(t) + \xi(t) \qquad (5.4.5)$$

This enlarged equation has a white noise disturbance $\xi(t)$. From the study of the least-squares method, we know that consistent and efficient estimates of a_i°, b_i°, d_i° can be obtained by minimizing the loss function

$$V = \frac{1}{N} \sum_{t=1}^{N} \varepsilon(t)^2 = \frac{1}{N} \sum_{t=1}^{N} \left(D(q)[A(q)y(t) - B(q)u(t)] \right)^2 \qquad (5.4.6)$$

This implies that, in the identification a model should be used which has the same structure as the true process

$$D(q)A(q)y(t) = D(q)B(q)u(t) + \varepsilon(t) \qquad (5.4.7)$$

where $\varepsilon(t)$ is the residual; see Fig. 5.4.1.

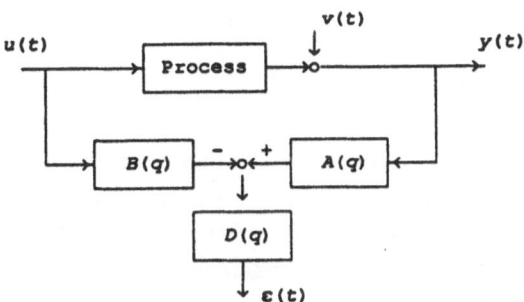

Fig. 5.4.1 Error generation of generalized least-squares method

We know that the minimization of loss function (5.4.6) has no analytical solution because the error $\varepsilon(t)$ is nonlinear in the parameters. We note, however, that the error $\varepsilon(t)$ of (5.4.7) has a special feature. For given $D(q)$ it is linear in $A(q)$ and $B(q)$, and vice versa. This feature can be exploited to obtain a simple algorithm for minimizing the loss function (5.4.6). Specifically, the algorithm consists of repeating the following two steps until convergence.

At iteration $k+1$:

Step 1 For given $\hat{D}^k(q)$ define the residual

$$\varepsilon_1^{k+1}(t) = A(q)[\hat{D}^k(q)y(t)] - B(q)[\hat{D}^k(q)u(t)]$$

The error $\varepsilon^k(t)$ is linear in $A(q)$ and $B(q)$, hence we can determine $\hat{A}^{k+1}(q)$ and $\hat{B}^{k+1}(q)$ by solving a LS problem where the loss function

$$V_1 = \frac{1}{N} \sum_{t=1}^{N} \varepsilon^{k+1}(t)^2 = \frac{1}{N} \sum_{t=1}^{N} [A(q)\hat{D}^k(q)y(t) - B(q)\hat{D}^k(q)u(t)]^2$$

is minimized. The formula of the LS estimator is now well known, so we ask our readers to write down the formula of the estimates $\hat{A}^{k+1}(q)$ and $\hat{B}^{k+1}(q)$.

Step 2 For given $\hat{A}^{k+1}(q)$ and $\hat{B}^{k+1}(q)$, define the residual

$$\varepsilon_2^{k+1}(t) = D(q)([\hat{A}^{k+1}(q)y(t)] - [\hat{B}^{k+1}(q)u(t)])$$

Then determine $\hat{D}^{k+1}(q)$ by minimizing

$$V_2 = \frac{1}{N} \sum_{t=1}^{N} \varepsilon_2^{k+1}(t) = \frac{1}{N} \sum_{t=1}^{N} D(q)([\hat{A}^{k+1}(q)y(t)] - [\hat{B}^{k+1}(q)u(t)])$$

This is again an LS problem.

Thus each step of the algorithm solves an LS problem. This is why the name of generalized least-squares (GLS) is given to the algorithm. The iteration can be started with a normal LS estimation. Fig. 5.4.2 shows the block diagram of error generation for the GLS algorithm.

There are several ways to modify the GLS algorithm in order to simplify the computation or to speed up the convergence rate. The main idea of these modifications is first to apply the LS method on the model (5.4.7) with order $n + n_d$ in order to obtain consistent estimates of polynomials $D(q)A(q)$ and $D(q)B(q)$, then to perform some kind of model reduction to retrieve $A(q)$, $B(q)$ and $D(q)$; see , e.g., Söderström and Stoica (1989), and Hsia (1977).

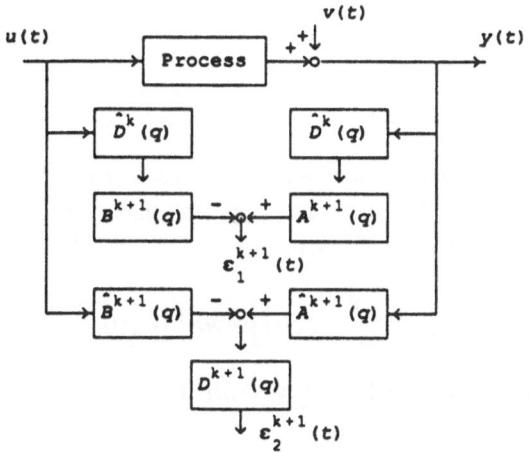

Fig. 5.4.2   Error generation of the GLS algorithm

For the problem of model validation and order selection, one can still use the simulation method. Now, however, there is another possibility. Because the GLS method aims at obtaining white noise residuals a natural way to do model validation and order selection is to check the whiteness of residuals for increasing orders. The sample autocorrelation function of the residuals can be used for this test. Note that we have to select both the process order and the order of the disturbance filter. To simplify the procedure, we can (in the first instance) let them be equal, i.e., $n = n_d$.

Now we shall explain why the GLS method can be called a prediction error method. Rewrite the true process (5.4.4) as

$$y(t) = \frac{B^\circ(q)}{A^\circ(q)}u(t) + \frac{1}{A^\circ(q)D^\circ(q)}\xi(t)$$

$$= \frac{B^\circ(q)}{A^\circ(q)}u(t) + \left[\frac{1}{A^\circ(q)D^\circ(q)} - 1\right]\xi(t) + \xi(t) \qquad (5.4.8)$$

Because the coefficients of the highest degree terms of $A^\circ(q)$ and $D^\circ(q)$ are 1 (they are monic polynomials), we know that their product will also have this property:

$$F^\circ(q) = A^\circ(q)D^\circ(q) = 1 + f_1 q^{-1} + \cdots + f_{2n} q^{-1}$$

Thus the filter

$$\left[\frac{1}{A^\circ(q)D^\circ(q)} - 1\right] = \frac{-f_1 q^{-1} - \cdots - f_{2n} q^{-1}}{A^\circ(q)D^\circ(q)}$$

has one unit delay. This means that the second term in (5.4.8) is a signal that only depends on the past data up to time t-1. When expressing this signal in terms of $u(t)$ and $y(t)$ we have

$$y(t) = \frac{B^\circ(q)}{A^\circ(q)} u(t) + \left[\frac{1}{A^\circ(q)D^\circ(q)} - 1\right] A^\circ(q)D^\circ(q)\left[y(t) - \frac{B^\circ(q)}{A^\circ(q)} u(t)\right] + \xi(t)$$

$$= (D^\circ(q)B^\circ(q)u(t) + [1 - D^\circ(q)A^\circ(q)]y(t)) + \xi(t)$$

$$= z(t) + \xi(t) \tag{5.4.9}$$

where

$$z(t) = (D^\circ(q)B^\circ(q)u(t) + [1 - D^\circ(q)A^\circ(q)]y(t))$$

Note that $z(t)$ and $\xi(t)$ are uncorrelated. If we use $z(t)$ as the one step ahead prediction of the the output $y(t)$, we have a prediction error $\xi(t)$ which is white noise. We would expect that this predictor is the best one in some sense, because when the prediction error is white noise, it contains no useful information at all. Indeed, we can show this more formally. Let $y^*(t)$ be an arbitrary predictor of $y(t)$. Then the variance of the prediction error is

$$E[y(t) - y^*(t)]^2 = E[z(t) + \xi(t) - y^*(t)]^2$$

$$= E[z(t) - y^*(t)]^2 + E\xi(t)^2 \geq E\xi(t)^2 = E[y(t) - z(t)]^2$$

Thus $z(t)$ is the optimal predictor in the sense of minimum variance.

In identification, we will write down the optimal filter in terms of unknown polynomials as

$$\hat{y}(t|\theta) = D(q)B(q)u(t) + [1 - D(q)A(q)]y(t) \tag{5.4.10}$$

and determine the parameters by minimizing the sum of the squares of the prediction errors:

$$V = \frac{1}{N} \sum_{t=1}^{N} [y(t) - \hat{y}(t)]^2 = \frac{1}{N} \sum_{t=1}^{N} [D(q)A(q)y(t) - D(q)B(q)u(t)]^2 \tag{5.4.11}$$

Again this is the loss function of the GLS method (5.4.6).

5.4.2 General Properties of the Family of Prediction Error Methods

The model structure (5.4.4) is one way to model the equation disturbance. Of course, other model structures can also be used. We will show more examples of model structures and then give some general discussions on prediction error methods.

ARMAX Model The true process model is assumed to be

$$A^\circ(q)y(t) = B^\circ(q)u(t) + C^\circ(q)\xi(t) \tag{5.4.12}$$

or

$$y(t) = \frac{B^\circ(q)}{A^\circ(q)}u(t) + \frac{C^\circ(q)}{A^\circ(q)}\xi(t)$$

where

$$\begin{cases} A^\circ(q) = 1 + a_1^\circ q^{-1} + \cdots + a_n^\circ q^{-n} \\ B^\circ(q) = b_1^\circ q^{-1} + \cdots + b_n^\circ q^{-n} \end{cases}, \quad C^\circ(q) = 1 + c_1^\circ q^{-1} + \cdots + c_{n_c}^\circ q^{-n_c}$$

and $\xi(t)$ is white noise with zero mean and variance R. So the equation disturbance is assumed to be an MA (moving-average) process. This model was proposed by Åström and Bohlin (1965) and has become a well known model. Similarly as for the GLS method, we can show that the predictor for this model structure has the form

$$\hat{y}(t|\theta) = \frac{B(q)}{C(q)}u(t) + \left[1 - \frac{A(q)}{C(q)}\right]y(t) \tag{5.4.13}$$

The parameters of this model are determined by minimizing

$$V = \frac{1}{N}\sum_{t=1}^{N}\left(\frac{1}{C(q)}[A(q)y(t) - B(q)u(t)]\right)^2 \tag{5.4.14}$$

An algorithm for this problem can be derived using the quasilinearization technique used for the output error method in Section 5.2. In identification literature this method is also called maximum likelihood method.

When comparing the GLS method and ARMAX model method, the advantage of the GLS method, as we have seen, is that simple minimization algorithms can be derived for the model structure; the advantage of ARMAX model is that it is very suitable for some controller design techniques; see Åström and Wittenmark (1984).

Note that the common idea of both GLS method and ARMAX model method is to model the equation disturbance; the consequence of this is that the process transfer function and output disturbance share the same $A(q)$ polynomial.

Modelling the Output Disturbance—Box-Jenkins Method

A natural development of the output error method is to further model the output disturbance $v(t)$. This can be done by assuming that the true process is

$$y(t) = \frac{B^{\circ}(q)}{A^{\circ}(q)}u(t) + \frac{C^{\circ}(q)}{D^{\circ}(q)}\xi(t) \tag{5.4.15}$$

where

$$\begin{cases} A^{\circ}(q) = 1 + a^{\circ}q^{-1} + \cdots + a_n^{\circ} q^{-n} \\ B^{\circ}(q) = \qquad b_1^{\circ}q^{-1} + \cdots + b_n^{\circ} q^{-n} \end{cases} \quad \begin{cases} C^{\circ}(q) = 1 + c^{\circ}q^{-1} + \cdots + c_n^{\circ} q^{-n} \\ D^{\circ}(q) = 1 + d_1^{\circ}q^{-1} + \cdots + d_n^{\circ} q^{-n} \end{cases}$$

This model structure was suggested by Box and Jenkins (1970). The predictor for this model is

$$\hat{y}(t|\theta) = \frac{D(q)B(q)}{C(q)A(q)}u(t) + \frac{C(q) - D(q)}{C(q)}y(t) \tag{5.4.16}$$

The parameters of this model are determined by minimizing

$$V = \frac{1}{N} \sum_{t=1}^{N} \left(\frac{D(q)}{C(q)}[y(t) - \frac{B(q)}{A(q)}u(t)] \right)^2 \tag{5.4.17}$$

The Box-Jenkins model has several advantages over the output error method. Firstly, it will supply both a process model and a disturbance model. Soon it will be shown that the Box-Jenkins model will be consistent also in closed loop operation; this implies that a Box-Jenkins method will give a more accurate process model than an output error method for a given process under closed-loop conditions. Under open-loop experiment, however, both methods will give consistent estimates. One may then ask which model is more accurate. The answer is postponed for a while; the reader is encouraged to guess it, using the knowledge gained so far.

A General Model

The following model is used by Ljung (1987) to cover all the existing model structures. The true process is assumed to be

$$F^{\circ}(q)y(t) = \frac{B^{\circ}(q)}{A^{\circ}(q)}u(t) + \frac{C^{\circ}(q)}{D^{\circ}(q)}\xi(t)$$ (5.4.18)

or

$$y(t) = \frac{B^{\circ}(q)}{F^{\circ}(q)A^{\circ}(q)}u(t) + \frac{C^{\circ}(q)}{F^{\circ}(q)D^{\circ}(q)}\xi(t)$$

where $A^{\circ}(q)$, $B^{\circ}(q)$, $C^{\circ}(q)$, $D^{\circ}(q)$ and $F^{\circ}(q)$ are polynomials of q^{-1}, all being monic except $B(q)$, $\xi(t)$ is white noise with zero mean and variance R.

This structure is too general for practical applications. One would fix one or more polynomials to unity for a specific process. However, by developing algorithms and results for the general model, we also cover all the special cases. The predictor for (5.4.18) is

$$\hat{y}(t|\theta) = \frac{D(q)B(q)}{C(q)A(q)}u(t) + \left[1 - \frac{D(q)F(q)}{C(q)}\right]y(t)$$ (5.4.19)

Consistency Analysis

Assume that the true process satisfies

$$y(t) = G^{\circ}(q)u(t) + H^{\circ}(q)\xi(t)$$ (5.4.20)

where $G^{\circ}(q)$ is the stable process transfer function, $H^{\circ}(q)$ is the disturbance filter which is stable and minimum phase (its inverse is stable too), and $\xi(t)$ is white noise with zero mean and variance R. We further assume that

$$\begin{cases} G^{\circ}(0) = 0 & (G^{\circ}(q) \text{ has at least one delay}) \\ H^{\circ}(0) = 1 & (H^{\circ}(q) \text{ is monic}) \end{cases}$$ (5.4.21)

Assume that the correct model structure is used. The predictor for the process is

$$y(t|\theta) = \frac{G(q,\theta)}{H(q,\theta)}u(t) + \left[1 - \frac{1}{H(q,\theta)}\right]y(t)$$ (5.4.22)

where θ is the parameter vector containing all the parameters of the process model and the disturbance model. The prediction error is

$$\varepsilon(t|\theta) = \frac{1}{H(q,\theta)}[y(t) - G(q,\theta)u(t)] \tag{5.4.23}$$

The parameter vector θ is determined by minimizing

$$V^N(\theta) = \frac{1}{N}\sum_{t=1}^{N}\varepsilon(t|\theta)^2 = \frac{1}{N}\sum_{t=1}^{N}(\frac{1}{H(q,\theta)}[y(t) - G(q,\theta)u(t)])^2 \tag{5.4.24}$$

When $N \to \infty$, we have

$$V^N(\theta) \to E\varepsilon(t|\theta)^2 = E(\frac{1}{H(q,\theta)}[y(t) - G(q,\theta)u(t)])^2 \tag{5.4.25}$$

It follows from (5.4.20) and (5.4.23) that

$$\varepsilon(t|\theta) = \frac{1}{H(q,\theta)}[G^\circ(q)u(t) + H^\circ(q)\xi(t) - G(q,\theta)u(t)]$$

$$= \frac{1}{H(q,\theta)}[G^\circ(q) - G(q,\theta)]u(t) + \frac{H^\circ(q)}{H(q,\theta)}\xi(t)$$

$$= \xi(t) + \text{a term independent of } \xi(t) \tag{5.4.26}$$

The last equality is due to the conditions in (5.4.21). Hence we have

$$V_\infty(\theta) = E\varepsilon(t|\theta)^2 \geq E\xi(t)^2 = R \tag{5.4.27}$$

and

$$E\varepsilon(t|\theta)^2 = \frac{1}{H(q,\theta)^2}\Delta G(q,\theta)^2 Eu(t)^2 + \frac{H^\circ(q)^2}{H(q,\theta)^2}R \tag{5.4.28}$$

If the global minimum is obtained for all N and if the input is persistently exciting with a sufficiently high order, then we can see from (5.4.28) that the lower bound of the asymptotic loss function (5.4.27) is attained when

$$\Delta G(q,\theta)^2 = 0, \quad \frac{H^\circ(q)^2}{H(q,\theta)^2} = 1$$

This implies that when $N \to \infty$

$$G(q,\theta) = G^\circ(q), \quad H(q,\theta) = H^\circ(q)$$

Therefore, the prediction error estimate is consistent. Note that we do not assume open-loop operation. It is essential that we should choose the error at the place where (assumedly) white noise enters the true process.

Remarks— When the process is operating in closed loop, both the order of the process model and the order of the disturbance model should be correct in order to get a consistent estimate. For open loop data, the above condition can be somewhat relaxed. If the process model and the disturbance model are parametrized independently, i.e., $F(q) = 1$ in (5.4.18), we can obtain a consistent estimate of the process model for open loop data even when the order of disturbance model is too low. We have shown that the output error model is consistent for an open loop experiment. The same can be done for the Box-Jenkins model. This property can be desirable in practice, because the accuracy of the process model is more important than that of the disturbance model.

— When the model structure is incorrect, then a prediction error model is biased. We can derive a asymptotic frequency domain expression for the loss function (5.4.25) for an open loop experiment (using Parseval's formula)

$$V_\infty = \frac{1}{2\pi}\int_{-\pi}^{\pi} (G^\circ(e^{i\omega}) - \frac{B(e^{i\omega})}{A(e^{i\omega})})^2 \frac{\Phi_u(\omega)}{H(e^{i\omega})^2} d\omega + \frac{1}{2\pi}\int_{-\pi}^{\pi} \frac{\Phi_v(\omega)}{H(e^{i\omega})^2} d\omega \qquad (5.4.29)$$

We see that the disturbance model is part of the frequency weighting for the process transfer function approximation. The expression for closed loop data is rather complex.

Relationship to the Maximum Likelihood (ML) Method

The *maximum likelihood* (ML) estimate of θ is obtained by maximizing the likelihood function, i.e., the probability density function of observations conditioned on the parameter vector θ. Suppose that the observations are represented by the random variable $y^T = [y(1), y(2), \cdots, y(N)]$. Denote the probability density function of y as

$$f(\theta; y_1, y_2, \cdots y_N) = f_y(\theta; y) \qquad (5.4.30)$$

Here the unknown parameter vector θ that describes properties of the observed variable will be estimated using the observations in y. If the observed value of y is y^*, then the maximum likelihood (ML) estimator is to select a estimate so that the observed event becomes "as likely as possible." That is, we determine θ such that

$$L(\theta) = f_y(\theta;y^*) \tag{5.4.31}$$

is maximized. $L(\theta)$ in (5.4.31) is called *likelihood function* which is a deterministic function of the parameter vector θ once the numerical value y^* is inserted. (In contrast, the probability density function (5.4.30) is a function of random variable y for fixed parameter vector θ.)

We see that the principle of ML method is simple and intuitively attactive. Therefore, we would expect that it has some nice properties. Indeed, it can be shown that ML method gives asymptotically (when $N \to \infty$) unbiased and efficient (minimum variance) estimates. In this sense the ML method gives the best possible estimator.

In order to apply the ML method to the identification of a linear process, we need to introduce the further assumption that the white noise in the true process (5.4.20) $\xi(t)$ is Gaussian (normally) distributed. It is equally valid to use the probability density function of the disturbance because there is a one-to-one transformation between $\{y(t)\}$ and $\{\xi(t)\}$ as given by (5.4.20) if the effect of initial conditions is neglected. Using the expression for the multivariable Gaussian distribution function, we have

$$L(\theta) = f_\xi(\theta,\underline{\varepsilon}) = \frac{1}{(2\pi)^{N/2} R^{1/2}} \exp\left[-\frac{1}{2} \sum_{t=1}^{N} \varepsilon(t,\theta)^2 / R \right] \tag{5.4.32}$$

Maximizing this likelihood function is equivalent to minimizing the loss function

$$-\ln L(\theta) = \frac{1}{2} \sum_{t=1}^{N} \varepsilon(t,\theta)^2 / R + \frac{1}{2} \ln R + C \text{ (constant)} \tag{5.4.33}$$

We find that (5.4.33) is equivalent to the loss function of the prediction error methods (5.4.24). Therefore, we can conclude that the prediction error methods are maximum likelihood method if the white noise $\xi(t)$ has a Gaussian distribution, hence give efficient estimates. This conclusion gives the prediction error methods a sound theoretical basis.

Now let us answer a question raised before: under open loop experiment, which model is more accurate, the output error model or the Box-Jenkins model? Assume that the output disturbance $v(t)$ is generated by filtering a white Gaussian noise, then a Box-Jenkins model with the right model orders of the process and the disturbance will be more accurate than a output error model. The reason is that the Box-Jenkins model is a ML estimate which is efficient and the output error model is not.

In practice, if the output disturbance consists only of measurement noise, the assumption that $v(t)$ is white Gaussian could be realistic; then the output error method is a ML method. A LS (least-squares) model is an ML estimate only if the equation disturbance $e(t)$ is white Gaussian, which is not at all a realistic assumption.

5.5 Identifying the Glass Tube Process Using a Prediction Error Method

Here the glass tube process is identified using an ARMAX model. Let us recall the definitions of input/output variables

$u_1(t)$: input 1, gas pressure; $\quad\quad$ $u_2(t)$: input 2, drawing speed;

$y_1(t)$: output 1, wall thickness; $\quad\quad$ $y_2(t)$: output 2, diameter.

We have shown that when using a diagonal form MFD (matrix fraction description) model, it is straightforward to generalize the SISO methods for MIMO processes. For the 2-input 2-output process, the diagonal form MFD model of the process is given as

$$G(q) = \begin{bmatrix} A_1(q) & 0 \\ 0 & A_2(q) \end{bmatrix}^{-1} \begin{bmatrix} B_{11}(q) & B_{12}(q) \\ B_{21}(q) & B_{22}(q) \end{bmatrix} \quad\quad (5.5.1)$$

Modelling the equation errors as MA (moving average) processes, we obtain an ARMAX model for the process:

$$\begin{cases} A_1(q)y_1(t) = B_{11}(q)u_1(t) + B_{12}(q)u_2(t) + C_1(q)\varepsilon_1(t) \\ A_2(q)y_2(t) = B_{21}(q)u_1(t) + B_{22}(q)u_2(t) + C_2(q)\varepsilon_2(t) \end{cases} \quad\quad (5.5.2)$$

where $A_i(q)$, $B_{ij}(q)$ and $C_i(q)$ are polynomials of the delay operator q^{-1}; and $A_i(q)$, and $C_i(q)$ are monic. Hence, we get two MISO sub-models; each of which can be estimated separately. For the simplicity, we let the degrees of all the polynomials of each sub-model be equal and denote this degree as n_i. Then the indices $[n_1, n_2]$ will determine the structure of the model.

For the ARMAX model estimation and validation, the data which have been used for the LS method are used here again, namely, the data from PRBS experiment. Again the first 600 samples are used for model estimation and the remaining 669 samples for model validation. For easy comparison, the model structure [4, 4] is used. Fig. 5.5.1 shows the result of model validation. The relative simulation errors of the two outputs are 45.4% and

14.4% respectively. We find that the ARMAX model brings a significant improvement over the LS model for the second sub-model; whereas the accuracy of the first sub-model remains very low. Possible reasons for this are that the minimization for the first sub-model is stuck at a local minimum, or the order of the sub-model is too low for a good approximation.

What we can learn from this example is that a prediction error model can indeed be more accurate than an LS model; on the other hand, the problems of local minima and undermodelling can be severe when noisy data from an industrial process are used. In Chapters 6 and 7, we will identify this process using our newly developed method and more comparisons will be made. There we will see that the problem with the ARMAX model is caused by undermodelling.

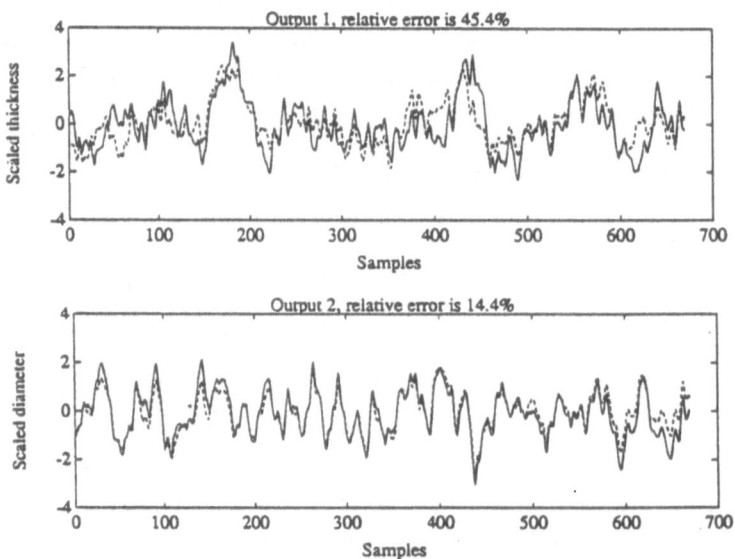

Fig. 5.5.1   Validation of the ARMAX model of the glass tube process
"solid" := measured output; "dashed" := model output

5.6   Conclusions and Discussion

In this chapter, we have studied different ways of modifying the least-squares method in order to overcome the bias problem. Looking at the model misfit in the frequency domain, we have rederived the Steiglitz-McBride method which is simple to understand and easy to use. The output error method has been motivated by choosing the identification criterion in a natural way. Using the correlation technique, we have obtained the instrumental variable (IV) methods. The family of prediction error methods have been motivated in order to obtain white noise residuals.

We have shown that most methods studied will give consistent estimates under weak conditions. The prediction error methods can give consistent estimates also for closed loop experiments; and the accuracies of the prediction models are generally higher than those of the output error model and of the IV models. The price for these advantages of the prediction error methods is more extensive computation. When the model order is incorrect (undermodelling), we have given, for most of the methods, transfer function error distributions in the frequency domain.

In our treatment, we have focused on the essence of the ideas that motivated the different methods. We have increased the level of abstraction very gradually. This process reflects, in some degree, the real development of identification techniques and theory in the last two decades. Hence the study of this chapter can be very helpful in understanding the vast literature of identification. In our presentation, we regard the prediction error methods and the maximum likelihood method as guiding principles rather than concrete methods. The understanding of these principles is useful in research and application of identification. However, most of the methods can be presented without using these more theoretical concepts. So if a reader does not have enough time to digest all the abstract theory, he is still able to use some relative simple methods from the first half of this chapter. For theoretically more rigorous treatments of these identification methods, we recommend the books of Ljung (1987) and Söderström and Stoica (1989).

For the choice among the different methods, if the user is only interested in process transfer function and if the process is operating in open loop, we would suggest to use the output error method (or even Steiglitz-McBride method). If very high accuracy is required, or also a disturbance model is needed, or if the process is operating in a closed loop, we suggest the use of Box-Jenkins method. If some physical *a priori* knowledge about the model structure is available, one should of course try to make use of it in choosing the identification method (or equivalently the model structure).

The theory of the prediction error methods has been considered as a mature part in the field of identification. However, when applying the theory to model industrial processes, the following difficulties exist:

- Numerical difficulties. As mentioned before, in general there exists no algorithm that guarantees finding the location of the global minimum. Some time ago, the cost of computing capacity was also a problem but this is now of less concern due to the drastic increase of inexpensive computing power.

- It is difficult to perform model order/structure selection for MIMO processes. It is extremely tedious to try all the different combinations of model order/structure in order to search for the correct one. Moreover, the problem of local minima will hinder this process.

- It is difficult to perform optimal input design for intended model applications such as control.

- It is difficult to supply a description (quantification) of model error of this model class that is suitable for robust control system analysis and design. This topic is of considerable interest and there is much on going research.

In order to overcome these problems of prediction error methods, in the next few chapters we will develop an identification method which can not only supply an accurate process model and disturbance model, but also give upper bounds of the model errors in the frequency domain. The methods are numerically simpler and more reliable than the prediction error methods. The problems of order/structure selection can be solved in a straightforward way; (nearly) optimal input design becomes an easy task.

CHAPTER 6

MIMO PROCESS IDENTIFICATION: A MARKOV PARAMETER APPROACH

In the previous two chapters, we have studied the LS method and different extensions of the LS method; most of the extensions belong to the family of the prediction error methods. We have seen that the properties of the LS method and those of the prediction error methods are complementary: the LS method is numerically simple and reliable, but model quality is low due to the bias of the estimate; the prediction error methods can give accurate estimates, but the algorithms are numerically difficult. Logically, one may think of combining the advantages of the LS method and the prediction error methods in some way. Moreover, there are common shortcomings of these methods: for example, it is difficult to determine model structure for a MIMO process. In this chapter we shall present an identification method which is suitable for MIMO process identification and numerically reliable. The method is developed by Backx and coworkers (see Backx, 1987, Backx and Damen, 1989).

This method is motivated in Section 6.1; and described in Section 6.2. In Section 6.3 the method is used to identify the glass tube manufacturing process.

6.1 Rationale of the Method

The purpose of this method is to estimate compact and accurate models of a large class of industrial processes for model based control or internal model control systems; see Backx and Damen (1989). In this control scheme the model is placed in parallel to the process and it simulates the process online.

As discussed in Chapter 5, if the model is used for simulation the output error criterion is a good choice. Hence this criterion is chosen for the method. In Chapter 5 we have learnt that the output error is nonlinear in the $A(q)$ parameters or the autoregressive parameters except for the FIR model or Markov parameter model, because $A(q) \equiv I$ in this model structure. Hence the equation error criterion and the output error criterion are the same for FIR models. This fact motivates us to begin with the estimation of a FIR model with a sufficient length.

The drawback of the FIR model, however, is the large number of parameters to be estimated and hence a high order of the model. This will results in large variances of the frequency responses of the model; see Chapter 7. Also a high order model is not suitable for controller design and implementation. Therefore model order should be reduced in order to reduce the variance of the model and to obtain a compact model that is suitable for controller design.

The question which immediately follows is what form should the reduced model have. Most literature on theoretical aspects of this topic advocates starting with canonical state space representations or MFD models of low McMillan degree and structural or Kronecker indices. The drawbacks related to the use of canonical forms is the structure selection procedure which is very complicated for a MIMO process. There is a way out when we use the so called MPSSM (minimum polynomial and start sequence of Markov parameters).

Let us first introduce the MPSSM model set. Given a process with state space model of order (McMillan degree) δ:

$$x(t+1) = Ax(t) + Bu(t)$$
$$y(t) = Cx(t) + Du(t) \tag{6.1.1}$$

where $x(t) = [x_1(t) \cdots x_\delta(t)]^T$ is a state space vector; $u(t)$ is the input vector and $y(t)$ is the output vector. From this model the input/output relation can be written as:

$$y(t) = CA^t x_0 + \sum_{k=1}^{t} CA^{k-1}Bu(t-k) + Du(t) \tag{6.1.2}$$

where x_0 is the initial state of the process. Assuming that initial state is zero, the Markov parameters of the process are obtained from (2.2.5) and (6.1.2)

$$G_k = \begin{cases} D = 0 & k = 0 \\ CA^{k-1}B & k > 0 \end{cases} \tag{6.1.3}$$

Note that we let $D = 0$ because the existance of time delays. The Cayley-Hamilton theorem states that every square matrix satisfies its characteristic equation (see Kailath, 1980). Application of the theorem to the state matrix A gives:

$$f(A) = A^{\delta} + \alpha_1 A^{\delta-1} + \cdots + \alpha_{\delta-1} A + \alpha_{\delta} I = 0 \qquad (6.1.4)$$

where $f(A)$ is called characteristic polynomial of A. Multiplication of equation (6.1.4) with some power of the state matrix A leads to the recurrent relation that can be used to compute any power of A:

$$A^{k-1} = \sum_{l=1}^{\delta} -\alpha_l A^{k-l-1} \qquad (6.1.5)$$

Left multiplication of equation (6.1.5) with the output matrix C and right multiplication with the input matrix B results in a recurrent relation for the Markov parameters:

$$G_k = \sum_{l=1}^{\delta} -\alpha_l G(k-l) \qquad \text{for all } k > \delta \qquad (6.1.5)$$

In general, however, the characteristic polynomial is not necessarily the polynomial of the lowest degree with the property given by (6.1.4). It can be shown (see Gantmacher, 1957) that the minimum degree r with this property is equal to the number of distinct eigenvalues of the state matrix A. This gives:

$$f_{mp}(A) = A^r + \alpha_1 A^{r-1} + \cdots + \alpha_{r-1} A + \alpha_r I = 0 \qquad r \le \delta \qquad (6.1.6)$$

Thus the recurrent relation for the Markov parameters can be reduced to

$$G_k = \sum_{l=1}^{r} -\alpha_l G(k-l) \qquad \text{for all } k > r \qquad (6.1.7)$$

Finally the input/output relation under the MPSSM model structure is given as:

$$y(t) = \sum_{k=0}^{t} G_k u(t-k) + v(t) \qquad (6.1.8)$$

with

$$G_k = \begin{cases} D = 0 & k = 0 \\ G_k & k = 1, 2, \cdots, r \\ \sum_{l=1}^{r} -\alpha_l G_k & k > r \end{cases} \qquad (6.1.9)$$

So the structure of the MPSSM model is determined by the degree of the minimum polynomial r, and this degree can be determined by examining the singular values of the block Hankel matrix built up by the estimated Markov parameters. The total number of parameters of the MPSSM model are in general much smaller than that of the corresponding FIR model.

The McMillan degree of an estimated MPSSM model is generally $r \cdot \min(m,p)$ with m the number of inputs and p the number of outputs (see Backx, 1987). This McMillan degree is larger than the order of a comparable minimal state space canonical form, but this problem can be reduced by applying a model reduction to the MPSSM model.

We shall estimate the parameters of the MPSSM model from the input/ output data and using an output error criterion. Because the output error is nonlinear in the parameters of the MPSSM model, some nonlinear numerical optimization algorithm has to be used. In order to avoid numerical local minimum problem and speed up the minimization procedure, a good initial estimate of the parameters is essential. This initial estimate can be obtained by subsequent application of the following steps:

— Estimate a FIR model from the data.

— Fit a MPSSM model to this FIR model by some model reduction technique.

The MPSSM model obtained in the last step can then be used as an initial estimate in the final minimization procedure which fits an MPSSM model to the input/output data according to an output error criterion.

6.2 The Identification Procedure

Step 1 Estimation of a FIR model

The length of the FIR model n can be determined from the staircase or the step responses measurements as described in Chapter 3. If n is chosen three times the largest relevant time constant, the tail effect of the FIR model is negligible.

The least-squares estimate of FIR parameters has been described in Chapter 4. We have seen there that if n is sufficiently large and if the test inputs are persistent exciting with order n and independent of the output disturbances (open loop experiment), the FIR model is consistent.

Step 2   Model reduction of the FIR model to an MPSSM model

Denote the Markov parameters of the estimated FIR model as

$$\hat{G}_k, \ k = 1, 2, \cdots, n \qquad (6.2.1)$$

Then construct a block Hankel matrix as

$$\hat{\mathcal{H}}_b = \begin{bmatrix} col[\hat{G}_1] & col[\hat{G}_2] & \cdots & col[\hat{G}_l] \\ col[\hat{G}_2] & col[\hat{G}_3] & \cdots & col[\hat{G}_{l+1}] \\ \vdots & \vdots & & \vdots \\ col[\hat{G}_j] & col[\hat{G}_{j+1}] & \cdots & col[\hat{G}_n] \end{bmatrix} \qquad (6.2.2)$$

where $col[G_k]$ indicates that all columns of G_k are put below one another into one column vector. Note that this is *not* the normal Hankel matrix defined in (2.3.9). The reason of using this special Hankel matrix is that we want to obtain the minimum polynomial degree r instead of the McMillan degree δ. Suppose that the degree of the minimum polynomial of the given process is r and that the Markov parameters are estimated perfectly, then it is easy to understand that the rank of the block Hankel matrix in (6.2.2) will be equal to r (it has r nonzero singular values), because, according to the recurrent relation (6.1.7), the $(r + 1)$th column of the matrix is a linear combination of previous r columns. This fact will be used to determine the degree r as follows. Perform singular value decomposition on the Hankel matrix which is filled with the estimated Markov parameters. The rank of this matrix will be higher than r due to the estimation errors. However, a sharp decrease of the singular value as a function of r or stabilization of its singular values on the noise level indicates the appropriate value of r.

For fitting an MPSSM model to the FIR model a method proposed by Gerth (1972) may be used. If we substitute the estimated Markov parameters $\hat{G}_k$ for G_k in equation (6.1.7), we can use these Markov parameters for a least-squares estimate of the minimal polynomial coefficients α_1, α_2, $\cdots \alpha_r$. More precisely we get:

$$\mathcal{G}\alpha = V + \mathbf{e} \qquad (6.2.3)$$

where

$$\mathcal{G} = \begin{bmatrix} col[\hat{G}_1] & col[\hat{G}_2] & \cdots & col[\hat{G}_r] \\ col[\hat{G}_2] & col[\hat{G}_3] & \cdots & col[\hat{G}_{r+1}] \\ \vdots & \vdots & & \vdots \\ col[\hat{G}_{n-r}] & col[\hat{G}_{n-r+1}] & \cdots & col[\hat{G}_{n-1}] \end{bmatrix} \qquad (6.2.4)$$

$$v^T = [col[\hat{G}_{r+1}]^T \ col[\hat{G}_{r+2}]^T \ \cdots \ col[\hat{G}_n]^T]^T \qquad (6.2.5)$$

$$\alpha = -[\alpha_1 \ \alpha_2 \ \cdots \ \alpha_r]^T \qquad (6.2.6)$$

and e is the error vector that accounts for the misfit. We know from Chapter 4 that the least-squares estimate of α that minimizes the loss function $e^T e$ is given by

$$\hat{\alpha} = [\mathcal{G}^T\mathcal{G}]^{-1}\mathcal{G}v \qquad (6.2.7)$$

If we take this estimate of the minimal polynomial coefficients as sufficiently accurate, the start sequence of Markov parameters G_k, $k = 1, 2, \cdots, r$, can be estimated analogously as follows. Denote

$$M_s^T = [col[G_1] \ col[G_2] \ \cdots \ col[G_r]]^T \qquad (6.2.8)$$

as the start sequence of Markov parameters to be estimated; let

$$\hat{M}^T = [col[\hat{G}_1] \ col[\hat{G}_2] \ \cdots \ col[\hat{G}_n]]^T \qquad (6.2.9)$$

contains the Markov parameters of the estimated FIR model; and denote

$$H^T = [I \ \mathcal{A}E_r \ \mathcal{A}^2E_r \ \cdots \ \mathcal{A}^{n-r}E_r] \qquad (6.2.10)$$

with

$$\mathcal{A} = \begin{bmatrix} 0 & \cdots & 0 & -\hat{\alpha}_r \\ 1 & & & \\ 0 & 1 & & \vdots \\ \vdots & & \ddots & \\ 0 & \cdots & 1 & -\hat{\alpha}_1 \end{bmatrix} \qquad (6.2.11)$$

and

$$E_r = [0 \ 0 \ \cdots \ 0 \ 1]^T \qquad (6.2.12)$$

From the recurrent relation (6.1.7) we have

$$HM_s = \hat{M} + e \qquad (6.2.13)$$

here **e** is again the error vector. Then the least-squares estimates of M_s is

$$\hat{M}_s = [H^T H]^{-1} H^T \hat{M} \qquad (6.2.14)$$

The MPSSM model given by (6.2.7) and (6.2.14) appears to be sufficiently accurate for use as an initial estimate. Other model reduction techniques can also be used to obtain an MPSSM model from the FIR model; see Backx (1987).

Step 3   Adjustment of an MPSSM model to the input/output data

The criterion to be minimized is the sum of the squares of the output errors. The output errors are linear in the start sequence of Markov parameters, G_k, $k = 1, 2, \cdots, r$, but nonlinear (in a polynomial form) in the minimum polynomial coefficients α_1, α_2, $\cdots$, α_r. This structural property can be used in the minimization procedure. Then, at each iteration, the nonlinear optimization is only used to find new values for the minimum polynomial coefficients and the update of the start sequence of Markov parameters can be obtained by the linear LS solution. This approach speeds up the iteration process considerably. Simulation studies have shown that the algorithm is 4 times faster than the same minimization algorithm that does not use the structural property of MPSSM models; the model quality remains the same (see Backx, 1987).

The estimated MPSSM model can be written in a state space representation. The state space form is preferred in control design and in model reduction. Assume that the number of inputs is greater than that of the outputs, $m \geq p$, which is often the case in a control application, then we can write the MPSSM model into a canonical observability form as follows; see Backx (1987).

$$\hat{A} = \text{diag}[\hat{A}_1 \ \hat{A}_2 \ \cdots \ \hat{A}_p] \qquad (6.2.15)$$

with $\hat{A}_1 = \hat{A}_2 = \cdots = \hat{A}_p$ and

$$\hat{A}_1 = \begin{bmatrix} 0 & 1 & 0 & \cdots & 0 \\ \vdots & & 0 & 1 & \vdots \\ \vdots & \vdots & & \ddots & \vdots \\ 0 & 0 & & \cdots & 1 \\ -\hat{\alpha}_r & -\hat{\alpha}_{r-1} & & \cdots & -\hat{\alpha}_1 \end{bmatrix}$$

$$\hat{C} = \begin{bmatrix} 1\ 0\ \cdots\ 0 & 0\ 0\ \cdots\ 0 & \cdots & 0\ 0\ \cdots\ 0 \\ 0\ 0\ \cdots\ 0 & 1\ 0\ \cdots\ 0 & \cdots & 0\ 0\ \cdots\ 0 \\ \vdots & & & \vdots \\ 0\ 0\ \cdots\ 0 & 0\ 0\ \cdots\ 0 & \cdots & 0\ 0\ \cdots\ 0 \\ 0\ 0\ \cdots\ 0 & 0\ 0\ \cdots\ 0 & \cdots & 1\ 0\ \cdots\ 0 \end{bmatrix} \tag{6.2.16}$$

$$\hat{B} = \begin{bmatrix} \hat{G}_{11,1} & \hat{G}_{12,1} & \cdots & \hat{G}_{1m,1} \\ \hat{G}_{11,2} & \hat{G}_{12,2} & \cdots & \hat{G}_{1m,2} \\ \vdots & \vdots & & \vdots \\ \hat{G}_{11,r} & \hat{G}_{12,r} & \cdots & \hat{G}_{1m,r} \\ \hat{G}_{21,1} & \hat{G}_{22,1} & \cdots & \hat{G}_{2m,1} \\ \vdots & \vdots & & \vdots \\ \hat{G}_{p1,1} & \hat{G}_{p2,1} & \cdots & \hat{G}_{pm,1} \\ \vdots & \vdots & & \vdots \\ \hat{G}_{p1,r} & \hat{G}_{p2,r} & \cdots & \hat{G}_{pm,r} \end{bmatrix} \tag{6.2.17}$$

$$\hat{D} = 0_{p\times m} \tag{6.2.18}$$

In most cases the order of this model can be reduced without causing too much errors, because a MPSSM model for a finite dimensional process is not minimal. The model reduction technique used is the so called balanced method; see Moore (1981) and Pernebo and Silverman (1982). In this method the state space model is first transformed to a so called *balanced* realization, then the least controllable and least observable states are eliminated from the model.

6.3 Identification of the Glass Tube Manufacturing Process

The glass tube process studied in Section 4.3 and Section 5.5 will be used again as a test case. We have seen that both the LS method and the ARMAX method with structure [4, 4] cannot deliver a good model. The same data will be used here; where the first 600 samples are used for model estimation and the remaining 669 samples for model validation.

A FIR model with length $n = 50$ is estimated. The block Hankel matrix is formed using the estimated Markov parameters as in (6.2.2). The singular values of the block Hankel matrix is plotted in Fig. 6.3.1. From this plot we decide to estimate a MPSSM model of order 7. Then the Gerth's algorithm is used to obtain an initial MPSSM model and the MPSSM model is estimated from the input/output data using the quasi-Newton algorithm. The order of the MPSSM model is $2 \times 7 = 14$. Balanced model reduction is used to reduced the order and finally a state space model of order 8 is obtained. The impulse responses of the FIR model, of the MPSSM model and of the final state space model are plotted in Fig. 6.3.2. We find that the impulse responses of these three models are consistent with each other. The relative errors of the MPSSM model are 9.5% on output 1 and 11.1% on output 2; the relative errors of the final state space model are 10.1% and 12.6%. Model validation of the final state space model is given in Fig. 6.3.3. This model has the same McMillan degree as that of the ARX model and ARMAX model of order [4, 4], but it has a much higher accuracy.

The models of the glass tube process identified with the Markov parameter method will be used for controller design; see Chapter 9.

This process will be identified again using the two-step method that will be developed in the coming two chapters.

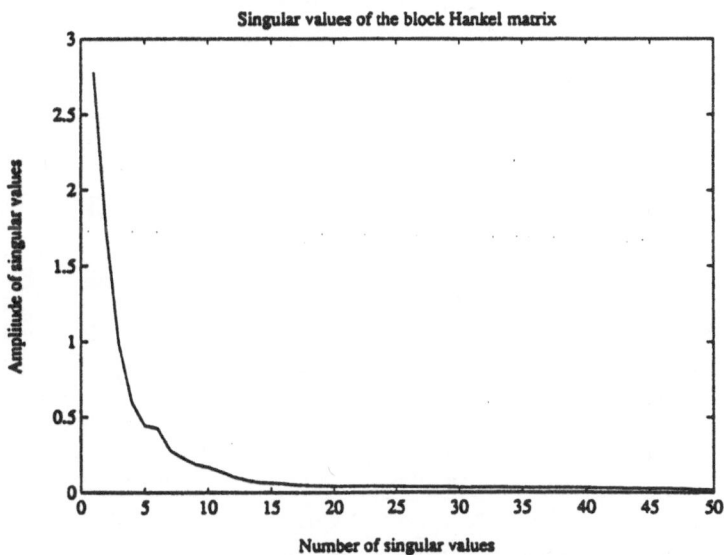

Fig. 6.3.1 Singular values of the block Hankel matrix

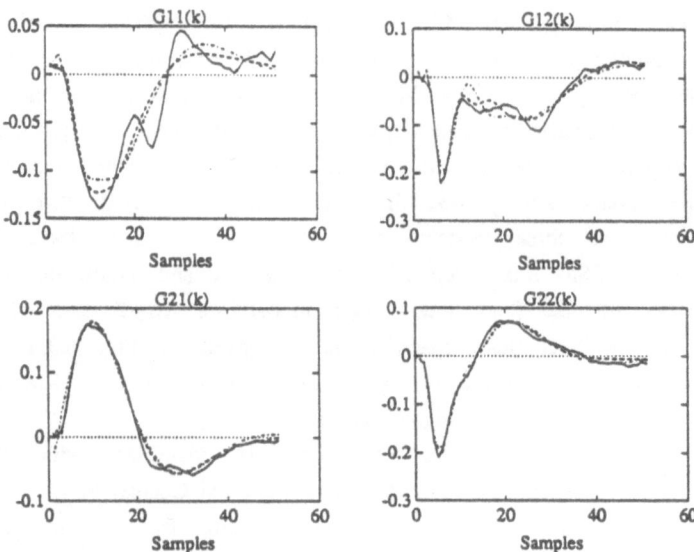

Fig. 6.3.2 Impulse responses of the FIR model (solid line), of the MPSSM model (dashed line) and of the final state space model (dashdot)

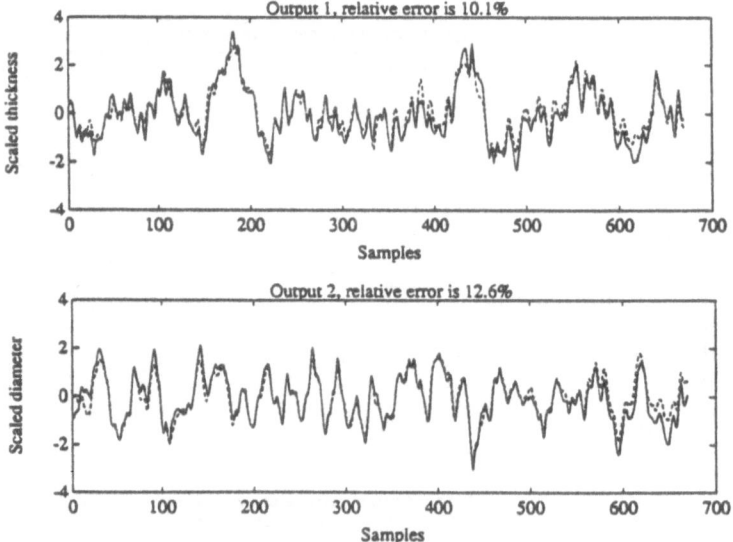

Fig. 6.3.3 Model validation of the 8th order state space model, "solid line": measured output, "dashed line": simulated output.

6.4 Conclusions

In this chapter we have presented an identification method that is suitable for MIMO process identification. The method starts with high dimensional FIR model estimation, a model reduction technique is used to obtain a MPSSM model, this MPSSM model is then used as the initial estimate for the refining of the model using the original input/output data. Finally the estimated MPSSM model is reduced to a state space model using a balanced model reduction method.

There are many advantages in using this method. Because it uses output error criterion, the estimated model will be consistent when the model structure contains that of the true process; it will give good approximation to the process frequency responses if the model structure is simpler than the true one; see Chapter 5. The method is numerically reliable due to the good initial estimate of the MPSSM model. In this approach, the difficult problem of MIMO model structure selection is avoided.

This method has been successfully applied to many MIMO industrial processes; compact and accurate models of these processes are obtained and multivariable model based control systems have been designed and implemented for them. The glass tube production process is one of the cases. For optimal disturbance reduction a model of the output disturbances is needed. This can be obtained by modelling the output error residuals as ARMA processes; see Ljung (1987) and Söderström and Stoica (1989) for modelling time series.

However, the method is not without limitations. Because it is an output error method, the estimate will be biased if the data are collected from a closed loop experiment. When the largest relevant time constant is much greater than the smallest one, the FIR model can have too many Markov parameters, this can cause problems in computer capacity. So far this method can not yet supply a quantitative description for the model errors that is needed for the analysis and design of robust controllers.

Thus more research work needs to be carried out.

CHAPTER 7

IDENTIFICATION FOR ROBUST CONTROL; SISO CASE

Because of our limited knowledge about reality, mathematical models can never give an exact description of the process behaviour under study. In the identification of industrial processes, undermodelling and disturbances are the main causes of model errors. In the previous decade, robust control theory has been proposed and developed; cf. Zames (1981), Doyle (1982), Vidyasagar (1985) and Morari and Zafiriou (1989). The advantage of robust control is its capability to cope with modelling errors in the analysis and design of control systems. In order to apply robust control theory, one needs not only a nominal process model, but also a suitable description of the modelling errors. These are typically in the form of some bounds on the model parameter variations of the parametric models; or the bounds on the frequency response variations (see Section 7.2).

Although the topic of linear process identification has been studied extensively over the past two decades, there are no straightforward methods which can be used to derive error bounds of identified models. Only very recently have some researchers started to study this problem; see, e.g., Goodwin and Salgado (1989), Kosut et al (1990), van den Boom et al (1991), Lamaire et al (1991) and Goodwin et al (1991). It seems that identification is not ready yet for its use in robust control, at least not for MIMO processes.

These observations have motivated us for the development of the method which will be presented in this chapter and the next one. The theme of the method is *identification for robust control.* Using this method, one can estimate not only an accurate parametric model of the process, but also an upper bound of the model errors in the frequency domain. The basic steps of the method consist of a high order model estimation and subsequent model reduction, which is similar to the Markov approach in Chapter 6. In this framework, fundamental problems such as *test input design* and *model structure selection* can easily be solved. The method is also numerically reliable. The method was developed by Zhu and co-workers; see Zhu (1990), Zhu and Backx (1991), Zhu, Backx and Eykhoff (1991).

In this chapter we will present the method for SISO processes; the MIMO generalization which is quite straightforward will be treated in the next chapter. First, in Section 7.1, the asymptotic theory of Ljung (1985) will be introduced which explains the frequency domain properties of prediction error methods when the model order increases. All the different steps of our method are based on this theory. In Section 7.2, the identification procedure is outlined and an upper bound of the model errors is defined. Then the subsequent two sub-sections are devoted to optimal input design and model order selection. In Section 7.3 we study recursive identification. A simulation example is presented in Section 7.4 and Section 7.5 contains the conclusions.

7.1 Asymptotic Properties of Prediction Error Models

Let us review the problem of linear process identification. Consider a SISO linear discrete-time process

$$y(t) = G^{\circ}(q)u(t) + H^{\circ}(q)\xi(t) \qquad (7.1.1)$$

where

$$G^{\circ}(q) = \sum_{k=1}^{\infty} g_k^{\circ} \cdot q^{-k} \text{ and } H^{\circ}(q) = \sum_{k=0}^{\infty} h_k^{\circ} \cdot q^{-k}$$

are called the transfer operators of the process and the disturbance respectively; q^{-1} is the unit delay operator; $y(t)$ and $u(t)$ are the output and the input at time t; $\{g_k^{\circ}\}$ is the impulse response of the process; $\{h_k^{\circ}\}$ is the impulse response of the disturbance filter, $\xi(t)$ is a zero mean white noise with variance R, and hence $v(t) = H^{\circ}(q)\xi(t)$ is a stationary stochastic process with zero mean value. The frequency response of the process is defined as

$$G^{\circ}(e^{i\omega}) = \sum_{k=1}^{\infty} g_k^{\circ} \cdot e^{-i\omega k} \qquad -\pi \le \omega \le \pi \qquad (7.1.2)$$

The identification problem is to estimate an approximate model from observed input/output data. Denote the data sequence by Z^N:

$$Z^N := y(1), u(1), \dots \dots, y(N), u(N) \qquad (7.1.3)$$

If we have a parametrized model (think of equation error model, ARMAX model, Box-Jenkins model or others):

$$y(t) = G(q,\theta)u(t) + H(q,\theta)\epsilon(t) \qquad (7.1.4)$$

where θ is the parameter vector, and $\epsilon(t)$ is white noise. Then a common way to determine the parameters in θ is to minimize the sum of the squared prediction errors

$$\frac{1}{N} \sum_{t=1}^{N} \varepsilon(t,\theta)^2 \tag{7.1.5}$$

where $\varepsilon(t,\theta)$ is the prediction error

$$\varepsilon(t,\theta) = y(t) - \hat{y}(t \mid \theta) = H^{-1}(q,\theta)[y(t) - G(q,\theta)u(t)] \tag{7.1.6}$$

As we have shown in Chapter 5, this problem statement can cover most of the time domain identification techniques in practice. Particular methods can be obtained by taking a specific model structure.

After the parameter estimation, the transfer function estimates of the process and the disturbance are denoted as:

$$\hat{G}_N^n(e^{i\omega}) = G(\hat{\theta},e^{i\omega}), \qquad \hat{H}_N^n(e^{i\omega}) = H(\hat{\theta},e^{i\omega}) \tag{7.1.7}$$

Here superscript n denotes the model order and the subscript N is to emphasize that N data samples are used for the estimation.

When the identified model is used in simulation and controller design, we are more concerned about the quality of the frequency response estimates than about the accuracy of parameters. The estimation of a frequency response is basically a non-parametric problem. Since the process is viewed as a black-box, the internal parametrization via θ is merely a vehicle to arrive at this estimate. Strictly speaking, many industrial processes are distributed parameter processes or, equivalently, infinite dimensional (order) processes; the glass tube process described in Section 4.3 is an example. Then, it is natural to let the model order n depend on the number of observed data samples, $n = n(N)$. Typically, in order to have a model set that is large enough to contain the true transfer function of an industrial process, or to give a good approximation of the true dynamics, we will allow the order n to increase when the number of data samples N increases, but n should be small compared to N. This can be formally expressed by:

$$n(N) \to \infty \text{ as } N \to \infty \quad \text{and} \quad n^2(N)/N \to 0 \text{ as } N \to \infty \tag{7.1.8}$$

For the identifiability of increasingly higher order models, we need that the process input should be persistently exciting with sufficiently high order. This can be expressed formally by assuming that the process input $u(t)$ is persistently exciting with any finite order, that is

$$\Phi_u(\omega) > 0 \quad \text{for } -\pi < \omega < \pi. \tag{7.1.9}$$

Result 7.1.1 (Ljung, 1985): Given (7.1.1)-(7.1.9). Suppose that the global minima are obtained in the minimizing the loss function (7.1.5) for all n and N. Then

$$\begin{bmatrix} \hat{G}_N^n(e^{i\omega}) \\ \hat{H}_N^n(e^{i\omega}) \end{bmatrix} \rightarrow \begin{bmatrix} G^o(e^{i\omega}) \\ H^o(e^{i\omega}) \end{bmatrix} \quad \text{w.p.1} \quad \text{as } N \rightarrow \infty \tag{7.1.10}$$

(Consistent estimates)

$$\sqrt{N} \begin{bmatrix} \hat{G}_N^n(e^{i\omega}) - E\hat{G}_N^n(e^{i\omega}) \\ \hat{H}_N^n(e^{i\omega}) - E\hat{H}_N^n(e^{i\omega}) \end{bmatrix} \rightarrow \mathcal{N}(0, P_n(\omega)) \quad \text{as } N \rightarrow \infty \tag{7.1.11}$$

(Asymptotic Gaussian distribution)

and

$$\lim_{n \to \infty} \frac{1}{n} P_n(\omega) = \Phi_v(\omega)\Phi^{-1}(\omega) \tag{7.1.12}$$

(Asymptotic covariance)

where

$$\Phi(\omega) = \begin{bmatrix} \Phi_u(\omega) & \Phi_{\xi u}(\omega) \\ \Phi_{u\xi}(\omega) & R \end{bmatrix}, \quad \Phi_v(\omega) = |H^o(e^{i\omega})|^2 R$$

and R is the variance of the white noise $\xi(t)$.

Implications of the result

These results show that the model estimates are consistent, and the errors of the transfer functions at each frequency follows a Gaussian distribution with a given covariance. The result is independent of the model structure; the reason for this is that the order is allowed to increase as the number of data points (experiment duration time) increases. Note that these results hold for closed loop experiment and they includes open-loop experiment as a special case. In order to be more generic, let us assume that the identification is performed under the feedback; see Fig. 7.1.1; where $K(q)$ denotes the feedback controller. We assume that the external test signal $w(t)$ is quasi-stationary (Ljung, 1987) meaning that its spectrum exists. Note that this test signal is *not* the process input $u(t)$.

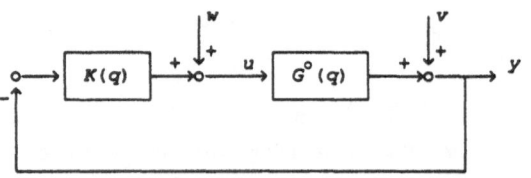

Fig. 7.1.1   Identification in a closed loop

Then from (7.1.12) we have an expression for the asymptotic variance of the process model

$$var[\hat{G}_N^n(e^{i\omega})] \approx \frac{n}{N} \frac{\Phi_v(\omega)R}{\Phi_u(\omega)R - |\Phi_{u\xi}(\omega)|^2}$$ (7.1.13)

For a clear physical insight, let us do a little bit more calculation. Denote

$$S(q) = \frac{1}{1 + G_o(q)K(q)}$$ (7.1.14)

as the *sensitivity* function (the transfer function from the disturbance $v(t)$ to the output $y(t)$). From the feedback relation

$$u(t) = w(t) - K(q)y(t)$$

we have

$$u(t) = S(q)w(t) - K(q)S(q)H_o(q)\xi(t)$$ (7.1.15)

Then from (7.1.15) we obtain

$$\Phi_u(\omega) = |S(e^{i\omega})|^2\Phi_w(\omega) + |K(e^{i\omega})|^2|S(e^{i\omega})|^2|H^o(e^{i\omega})|^2R$$

$$\Phi_{u\xi}(\omega) = K(e^{i\omega})S(e^{i\omega})H^o(e^{i\omega})R$$

Inserting these two equations into (7.1.13) we obtain

$$var[\hat{G}_N^n(e^{i\omega})] \approx \frac{n}{N} \frac{\Phi_v(\omega)}{\Phi_w(\omega)} \frac{1}{|S(e^{i\omega})|^2}$$ (7.1.16)

This expression is surprisingly simple! It say. that the variance of the model at a given frequency is proportional to the (output) noise-to-

(external test) signal ratio multiplied by the ratio of model order to the number of data samples, and is inversely proportional to the square of the sensitivity function (the effect of feedback). We see that in this theory, the model error (variance) is related to the disturbance, the input signal, the time duration of the experiment and the sensitivity function in a way that can be given a direct physical interpretation. Due to its simplicity, we can make extensive use of the theory in the development of our identification method.

For open loop experiment, we have

$$u(t) = w(t), \qquad S(q) = 1$$

Thus the variance expression (7.1.16) is further simplified to

$$var[\hat{G}_N^n(e^{i\omega})] \approx \frac{n}{N} \frac{\Phi_v(\omega)}{\Phi_u(\omega)} \qquad (7.1.17)$$

This can also be derived directly from (7.1.13) by letting $\Phi_{u\xi}(\omega) = 0$.

Comments on the establishment of Result 7.1.1

The mathematical proofs of Result 7.1.1 are quite involved; see Ljung (1985) and Ljung and Yuan (1985). But it is possible to give physical explanations to readily appreciate the validity of this theory.

It is not difficult to see why the estimates are consistent. As the order n increase, the model flexibility will increase. At some moment, the true structure is captured by the model and the consistency is obtained.

To see why the error of the frequency response has a Gaussian distribution with the given variance, let us assume, for the simplicity, that the process is operating in an open-loop, the disturbance is white noise with variance R and the input is (known) white noise with variance R_u, and a FIR model is estimated. Then based on the analysis in Section 4.4, we can show that the errors of the n parameter estimates are independent; and they all have the same variance $R/(NR_u)$. The frequency response estimate is the Fourier transform of the n FIR parameters; thus the Fourier transform of the FIR parameter errors are the errors of the transfer function estimate. What the Fourier transform does is simply to create a weighted summation of all the errors, where the modulus of each weight is 1. Therefore, applying the Central Limit Theorem to complex valued random variables, we know that when $n \to \infty$, the error of the transfer function follows a Gaussian distribution with a variance given by

$$\frac{n}{N}\frac{R}{R_u}$$

This is a special case of (7.1.17).

7.2 The Identification Method

In order to apply the asymptotic theory, our approach is to start with a high order model estimation and then to perform a model reduction to arrive at compact models. This approach has practical value. We can, for example, avoid numerical problems by combining high order estimation and model reduction. In this way one also can simplify the step of model order determination and structure selection, which is difficult for MIMO processes. Input design for a high order model is an easy task.

The asymptotic properties of the transfer function estimates are independent of model structures. So, it is sensible to use a model structure that is as simple as possible for the high order model. The equation error model is the most simple parametrization which can supply both the process model and disturbance model. The FIR model is even simpler, but the disturbance model is not estimated. Because analytical solutions exist for these two model structures, the global minimum is guaranteed for all n, and N; hence the asymptotic theory applies.

7.2.1 The Procedure

Step 1. High order model estimation

Estimate a high order, n in the range 20 to 40, equation error model; and denote the process model, disturbance model and the disturbance spectrum as

$$\hat{G}_N^n(q) = \frac{\hat{B}^n(q)}{\hat{A}^n(q)}, \quad \hat{H}_N^n(q) = \frac{1}{\hat{A}^n(q)}, \quad \hat{\Phi}_v(\omega) = \frac{\hat{R}}{|\hat{A}^n(e^{i\omega})|^2} \tag{7.2.1}$$

where $\hat{R}$ is the estimated variance of the equation error residual.

Step 2. Model reduction

The model that results from Step 1 is often over-parametrized meaning that the process danamics can be described or well approximated by a model with a much lower order. Since the variance is proportional to the order n, model reduction can reduce the variance if it is performed properly. Also it is more convenient to use a compact (reduced) model in controller design.

The asymptotic theory of Section 7.1 shows that, in the frequency domain, the high order model follows approximately a Gaussian distribution with the variance given by (7.1.16). If we view the frequency response of the high order estimates

$$\hat{G}^n(e^{i\omega_1}),\ \hat{G}^n(e^{i\omega_2}),\ \cdots\ \hat{G}^n(e^{i\omega_n}),\ \text{where}\ \omega_k = \frac{k\cdot\pi}{n},\ k = 1,\ \cdots,\ n$$

as the noisy observations of the true transfer function, we can then apply the maximum likelihood principle. It can be shown that when $n \to \infty$, the asymptotic negative log-likelihood function for the process model is given by (Wahlberg, 1989):

$$V = \frac{1}{2\pi} \int_{-\pi}^{\pi} |\hat{G}^n(e^{i\omega}) - \hat{G}^l(e^{i\omega})|^2\ \frac{\Phi_w(\omega)\ |S(e^{i\omega})|^2}{\Phi_v(\omega)}\ d\omega \tag{7.2.2}$$

Instead of reproducing the rather lengthy derivation, we can justify this result by observing the fact that the errors are weighted by the inverse of their variances (neglecting the factor N/n); this will lead to an asymptotically efficient (minimum variance) estimate of the frequency response. But now we also know that the high order model (observation) follows a Gaussian distribution, thus we could imagine that the reduced model is an asymptotic maximum likelihood estimate.

Solving this problem needs a nonlinear minimization algorithm and there are many methods to choose from. We will approach this problem by performing an output error identification method as follows:

1) **Simulation.** Collect the external inputs, $\{w(t),\ t = 1,\ \cdots, N\}$, which have been used in the identification experiment. Filter $w(t)$ by the inverse of the disturbance model and by the estimated sensitivity function

$$u_f(t) = \frac{1}{H^n(q)} \hat{S}(q)w(t) = \frac{\hat{A}^n(q)}{1 + \hat{G}_N^n(q)C(q)} w(t)$$

where $u_f(t)$ is used to realize the desired frequency weighting. Then simulate the high order model using the filtered input:

$$\hat{y}^n(t) = \frac{\hat{B}^n(q)}{\hat{A}^n(q)} [u_f(t)] \tag{7.2.3}$$

This is equivalent to

$$\hat{y}^n(t) = \hat{B}^n(q)\hat{S}(q)w(t) \tag{7.2.4}$$

which is simpler than (7.2.3). Thus we obtain the input/output data of the high order model $\hat{G}_N^n(q)$:

$$Z_n^N := [\hat{y}^n(1), \; u_f(1), \; \cdots, \; \hat{y}^n(N), \; u_f(N)]$$

2) **Parameter estimation** The parameters of the reduced model are calculated by using an *output error method* which minimizes the following loss function

$$V_{oe}^N = \frac{1}{N} \sum_{t=1}^{N} \left\{ [\hat{G}^n(q) - \hat{G}^l(q)] u_f(t) \right\}^2$$

Letting $N \to \infty$, and applying Parseval's identity we obtain

$$V_{oe}^\infty = \frac{1}{2\pi} \int_{-\pi}^{\pi} |\hat{G}_N^n(e^{i\omega}) - \hat{G}^l(e^{i\omega})|^2 \frac{\Phi_w(\omega)|\hat{S}(e^{i\omega})|^2}{|\hat{H}^n(e^{i\omega})|^2} \, d\omega \tag{7.2.5}$$

This is equivalent to the asymptotic likelihood function (7.2.2), except that the disturbance spectrum and the sensitivity function are replaced by their high order estimates.

In a similar way, a reduced disturbance model can be obtained from the high order estimate $1/\hat{A}^n(q)$. Then we have the reduced process model and the reduced disturbance model as

$$\hat{G}^l(q) = \frac{\hat{B}^l(q)}{\hat{A}^l(q)}, \quad \hat{H}^l(q) = \frac{\hat{C}^l(q)}{\hat{D}^l(q)} \tag{7.2.6}$$

Hence the final model has a Box-Jenkins structure.

Step 3. Deriving an upper bound of modelling errors

We know that the errors of the high order model follows asymptotically a normal distribution with the variance given by (7.1.16). Therefore a 3σ upper bound of the errors of the high order model can be define as follows:

$$|G^o(e^{i\omega}) - \hat{G}^n(e^{i\omega})| \leq 3\sqrt{\frac{n}{N}\frac{\Phi_v(\omega)}{\Phi_w(\omega)\,|S(e^{i\omega})|^2}} \qquad \text{w.p. } 99.99\% \qquad (7.2.7)$$

This upper bound can also be used for the reduced model $\hat{G}^l(q)$, because the model reduction will reduce the model error. Thus we have

$$|G^o(e^{i\omega}) - \hat{G}^l(e^{i\omega})| \leq \bar{\Delta}(\omega) := 3\sqrt{\frac{n}{N}\frac{\Phi_v(\omega)}{\Phi_w(\omega)\,|S(e^{i\omega})|^2}} \qquad \text{w.p. } 99.99\% \qquad (7.2.8)$$

where $\Phi_v(\omega)$ can be estimated by (7.2.1); $\Phi_w(\omega)$ can be calculated from the measurement, and $S(q)$ can be determined by

$$\hat{S}(q) = \frac{1}{1 + \hat{G}^l(q)K(q)}$$

We note that the asymptotic theory can also be used to estimate an upper bound for a given model $\hat{G}(q)$. To do this, perform an identification experiment and estimate a high order model $G_N^n(q)$ whose errors are bounded by (7.2.7). Therefore, via the high order model and its bound we obtain an upper bound for the given model

$$|G^o(e^{i\omega}) - \hat{G}(e^{i\omega})| \leq |G^o(e^{i\omega}) - \hat{G}^n(e^{i\omega})| + |\hat{G}^n(e^{i\omega}) - \hat{G}(e^{i\omega})|$$

$$\leq 3\sqrt{\frac{n}{N}\frac{\Phi_v(\omega)}{\Phi_w(\omega)\,|S(e^{i\omega})|^2}} + |\hat{G}^n(e^{i\omega}) - \hat{G}(e^{i\omega})| \qquad \text{w.p. } \geq 99.99\%$$

$$(7.2.9)$$

This is more conservative than (7.2.8) because the data have not been used for deriving the model.

Comments on the Proposed Method

The method can be called *two-step* approach. There are two reasons to start with a high order model estimation. The first one is to obtain a consistent estimation for the simple equation error (or LS) model; the second one is to make use of the asymptotic theory which assumes that the order tends to infinity. So, even when we know that the process has a low order, e.g., order 2, we will still use, in this step, order 20 or higher.

One may also use a FIR model in this step. This means we can supply an error upper bound estimate for the Markov parameter approach of Chapter 6. Then, for use in the next step, it is necessary to estimate a model of the disturbance. This can be done by estimating an AR model from the output error residuals by the LS method.

We have used the data set $\{y(t), u(t)\}$ for the estimation of the high order model, the data set $\{y(t), w(t)\}$ is used for model reduction. In practice all these signals can be obtained without difficulty.

There are two reasons that motivate us to use identification for model reduction: 1) We do not need to implement another minimization algorithm; we have already studied the output error method. 2) We can easily write our algorithm in a recursive form for online identification, which will be clear later.

Although we eventually also need a nonlinear minimization algorithm to find the reduced model, we are now in a much better position than if we use a prediction error method. First, the influence of the disturbance is reduced greatly when using the data from the simulation of the high order model instead of the original data. Secondly, we have many different ways to obtain initial values: LS estimate, Steiglitz-McBride estimate, or initial values obtained by some well known model reduction methods, e.g., (frequency weighted) balanced model reduction or (frequency weighted) Hankel-norm approximation. Thirdly, we can detect poor local minima. Because if the minimization algorithm converges to the global minimum, the frequency response of the reduced model should lie in the middle of the fluctuating frequency response of the high order model due to the smoothing effect of model reduction. If this is not the case, a local minimum is detected; and we should start a minimization with another initial estimate. All the three factors make our algorithm numerically highly reliable.

In the literature, an upper bound with a probability is called *soft bound*. The upper bound is a simple function of the design variables such as the spectrum of test signal, sensitivity function and experiment duration. This makes the interaction between identification and controller design possible. If the modelling errors at some frequencies are too large for control application, we know from (7.2.8) that we can reduce the errors by: (1) modifying the spectrum of the test signal $\Phi_w(\omega)$ (experiment design), (2) using more data (increase of experiment duration), and (3) modify the sensitivity function (use another controller during the experiment). However, we do not recommend to reduce n, the order of the high order model. A large n

is necessary for the consistency of the high order model and the validity of the asymptotic theory.

It should be clear that the whole procedure can also be applied for the data from an open loop experiment; this can be done by simply setting

$$w(t) = u(t), \quad S(q) = 1$$

7.2.2 Model Order Selection

The simulation method proposed in Chapter 4 can also be used for model validation and order selection for the reduced model obtained with our identification scheme. For our method, however, we will propose a method of order selection based on frequency domain measures.

The asymptotic ML model reduction will in general increase the model quality; if the reduced model is allowed to deviate from the high order model the same amount as the error of the high order model (measured by its variance), we are most hopeful that the reduced model is most close to the true frequency response. Based on this observation, our method of order selection is simply: choose the order such that the difference between the high order model and the reduced model (in the frequency domain) approximately equals the variance of the high order model. Therefore, the order of the reduced model is determined such that

$$|\hat{G}^n(e^{i\omega}) - \hat{G}^l(e^{i\omega})|^2 \approx \frac{n}{N} \frac{\hat{\Phi}_v(\omega)}{\hat{\Phi}_w(\omega)} \frac{1}{|\hat{S}(e^{i\omega})|^2} \tag{7.2.10}$$

This selection can be done simply by visual inspection. The same idea can be applied for determining the order of the disturbance model.

Remarks—In this selection rule, the selected order is usefully related to the noise-to-signal ratio, and to the experiment time. For a given process, if the noise level is high and the experiment time is short, the selected order of the reduced model will be low. For the same process, the selected order will increase if the power of the test signal and/or the experiment time increases

— There is a difference in concept between our method and many existing methods of order selection, e.g., the whiteness test of residuals. We do not intend to find the true order of the process. Instead, we search for an order so that the best frequency response estimate can be obtained. The true order can be found by our method if this order is much lower than n, the noise level is sufficiently low and/or if the experiment time is long.

7.2.3 Optimal Experiment Design for Simulation

In optimal experiment design, or optimal input design, one tries to derive a test signal that is optimal in some sense. Early research tried to use the input design in order to improve the accuracy of the parameters (see Mehra, 1974). There are two drawbacks to this approach. First, the optimization procedure for deriving the optimal input is in general very difficult. Secondly, the intended model application is not addressed in this approach.

The problem of experiment design should be related, as far as possible, to the intended use of the model, otherwise, it is only an academic exercise. Now let us assume that the identified model will be used in an *internal model control* scheme. It can be shown (Morari and Zafiriou, 1989) that the internal model control scheme is closely related to the Smith predictor control (well known to process control people) that is suitable for controlling an industrial process with large time delays. In this control scheme, the identified process model is placed in parallel with the process. The difference between the process output and simulated model output is fedback to the controller $Q(q)$. Therefore a model which can optimally replicate the underlying process behaviour for certain given inputs will be very suitable for a internal model control scheme.

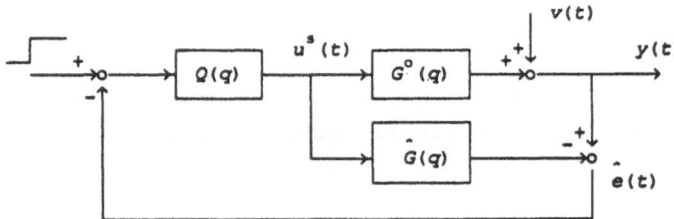

Fig. 7.2.1 The internal model control scheme

Denote $u^s(t)$ as the control input to the process (the simulation input to the model) with spectrum $\Phi_u^s(\omega)$. Let us now derive an optimal test signal

$w(t)$ for the high order model for the purpose of simulation. Using the high order model the simulation error is

$$e(t) = [\hat{G}_N^n(q) - G^o(q)]u^s(t) \tag{7.2.11}$$

Note that the disturbance is not included here, so it is not the output error defined before. The error in (7.2.11) has the variance (for the fixed $\hat{G}_N^n(q)$)

$$V_e = \frac{1}{2\pi}\int_{-\pi}^{\pi}|\hat{G}_N^n(e^{i\omega}) - G^o(e^{i\omega})|^2\Phi_u^s(\omega)d\omega \tag{7.2.12}$$

But now $\hat{G}_N^n(q)$ is a random variable due to disturbance. Hence we will choose the experimental conditions so that the averaged variance

$$EV_e \approx \frac{1}{2\pi}\int_{-\pi}^{\pi} \text{var}[\hat{G}_N^n(e^{i\omega})]\Phi_u^s(\omega)d\omega \tag{7.2.13}$$

is minimized. According to the asymptotic variance expression (7.1.16) it follows that

$$EV_e \approx \frac{1}{2\pi}\int_{-\pi}^{\pi} \frac{n}{N} \frac{\Phi_v(\omega)}{\Phi_w(\omega)\,|\,S(e^{i\omega})\,|^2}\Phi_u^s(\omega)d\omega \tag{7.2.14}$$

Applying the result of Yuan and Ljung (1985), we can show that the optimal spectrum of $w(t)$ which minimizes the mean square of the simulation error for the high order model is given by

$$\Phi_w^{op}(\omega) = \mu\sqrt{\Phi_u^s(\omega)\Phi_v(\omega)}\Big/|S(e^{i\omega})| \tag{7.2.15}$$

where μ is a constant which is adjusted with respect to the constraint of input amplitude.

For open-loop experiments, the optimal input spectrum is

$$\Phi_u^{op}(\omega) = \mu\sqrt{\Phi_u^s(\omega)\Phi_v(\omega)} \tag{7.2.16}$$

Notice that this spectrum is *not* related to the process transfer function. Result (7.2.16) says that more signal power should be put at those frequencies where the model will be used more intensively (simulation input has more power) and where more disturbance appears.

In practice we do not know the sensitivity $S(q)$ before identifying the process; and we do not know the signal $u^s(t)$ before designing the controller and testing the closed loop control system. This means that some iteration between identification experiment and control experiment is necessary for the optimal input design and finally optimal control. This is true also when other identification methods are used.

However, we can do nearly optimal input design before any identification and controller test as follows. Assume that the goal of the control is disturbance attenuation and setpoint tracking, and the identification experiment will be in an open-loop. Examine the feedback control system in Fig. 7.2.1, the control input is the filtered disturbance plus the step setpoint

$$u^s(t) = T(q)[v(t) + \beta 1(t)] \qquad (7.2.17)$$

where $T(q)$ is the unknown transfer operator from the disturbance and the step setpoint $1(t)$ to the input, β is a constant which is used to adjust the relative importance between disturbance reduction and setpoint tracking. The optimal input is according to (7.2.16)

$$\Phi_u^{op}(\omega) = \mu |T(e^{i\omega})| \sqrt{\Phi_v(\omega)[\Phi_v(\omega) + \beta^2 \Phi_{sp}(\omega)]} \qquad (7.2.18)$$

Because $T(q)$ is unknown, we can simply let $T(q) = C$ (constant). Then we have an approximation to the optimal input:

$$\Phi_u^{op}(\omega) = \mu C \sqrt{\Phi_v(\omega)[\Phi_v(\omega) + \beta^2 \Phi_{sp}(\omega)]} \qquad (7.2.19)$$

If $|T(e^{i\omega})|$ is flat over the low and middle frequency range, this formula will be a very good approximation. Hence, the nearly optimal input spectrum can be derived without knowing the process model and the controller; the output disturbance $v(t)$ can be measured from the uncontrolled process when keeping the input constant (free-run experiment). The signal with the desired spectrum can be generated by filtering a white noise signal. Now the method becomes surprisingly simple and economical.

7.3 Recursive Estimation

In many cases it may be necessary to estimate a model online while the process is in operation. The model will be updated when the new observations are available. Hence for computing efficiency, it is desirable to arrange the algorithms in such a way that the results obtained previously can be used for online updating. This way of computing the estimates is called recursive, or online, estimation which will be studied in this section. First we will derive the recursive LS algorithm and a recursive output error algorithm, then we show how to write our method into a recursive form.

7.3.1 The Recursive LS Method

Using the LS method to estimate an nth order transfer function model, we write down the linear regression

$$y(t) = \varphi(t)\theta + \varepsilon(t) \tag{7.3.1}$$

where

$$\varphi(t) = [-y(t-1) \cdots -y(t-n) \; u(t-1) \cdots u(t-n)]$$

$$\theta = [a_1 \cdots a_n \; b_1 \cdots b_n]^T$$

and $\varepsilon(t)$ is the equation error. The LS estimate which minimizes the loss function

$$V_{LS}(t) = \frac{1}{t}\sum_{k=1}^{t} \varepsilon(k)^2 = \frac{1}{t}\sum_{k=1}^{t} [y(k) - \varphi(k)\theta]^2$$

is

$$\hat{\theta}(t) = \left[\sum_{k=1}^{t} \varphi(k)^T\varphi(k)\right]^{-1}\left[\sum_{k=1}^{t} \varphi(k)^T y(k)\right] \tag{7.3.2}$$

The argument t is used here to indicate the dependence of $\hat{\theta}$ on time. Introduce the matrix

$$P(t) = \left[\sum_{k=1}^{t} \varphi(k)^T\varphi(k)\right]^{-1} \tag{7.3.3}$$

We have

$$\hat{\theta}(t) = P(t)\left[\sum_{k=1}^{t} \varphi(k)^T y(k)\right] \tag{7.3.4}$$

It is easy to see that

$$P(t+1)^{-1} = \left[\sum_{k=1}^{t+1} \varphi(k)^T\varphi(k)\right] = \left[\sum_{k=1}^{t} \varphi(k)^T\varphi(k)\right] + \varphi(t+1)^T\varphi(t+1)$$

$$= P(t)^{-1} + \varphi(t+1)^T\varphi(t+1) \tag{7.3.5}$$

Then at time $t+1$

$$\hat{\theta}(t+1) = P(t+1)\left[\sum_{k=1}^{t} \varphi(k)^T y(k) + \varphi(t+1)^T y(t+1)\right]$$

$$= P(t+1)\left[[P(t)^{-1}P(t)]\sum_{k=1}^{t} \varphi(k)^T y(k) + \varphi(t+1)^T y(t+1)\right]$$

[use (7.3.4)]

$$= P(t+1)[P(t)^{-1}\hat{\theta}(t) + \varphi(t+1)^T y(t+1)]$$

[use (7.3.5)]

$$= P(t+1)\{[P(t+1)^{-1} - \varphi(t+1)^T\varphi(t+1)]\hat{\theta}(t) + \varphi(t+1)^T y(t+1)]$$

$$= \hat{\theta}(t) + P(t+1)\varphi(t+1)^T[y(t+1) - \varphi(t+1)\hat{\theta}(t)] \qquad (7.3.6)$$

Hence we get the first version of the recursive formulas

$$\hat{\theta}(t+1) = \hat{\theta}(t) + K(t+1)\varepsilon(t+1) \qquad (7.3.7a)$$

$$\varepsilon(t+1) = y(t+1) - \varphi(t+1)\hat{\theta}(t) \qquad (7.3.7b)$$

$$K(t+1) = P(t+1)\varphi(t+1)^T \qquad (7.3.7c)$$

We know that the equation error $\varepsilon(t)$ can be interpreted as a prediction error. The interpretation of the recursive formula (7.3.7a) is, if $\varepsilon(t)$ is small, then estimate $\hat{\theta}(t)$ is good and should not be modified very much. The vector $K(t+1)$ is a weighting showing how much each parameter will be modified.

In this version (7.3.5) must be used to calculate $P(t+1)$. This needs a matrix inversion at each time step, which is very time-consuming. Using the matrix inversion lemma (where A and C are assumed invertible):

$$[A + BCD]^{-1} = A^{-1} - A^{-1}B[C^{-1} + DA^{-1}B]^{-1}DA^{-1}$$

we have

$$P(t+1) = \left[\sum_{s=1}^{t} \varphi(s)^T\varphi(s) + \varphi(t+1)^T\varphi(t+1)\right]^{-1}$$

$$= P(t) - P(t)\varphi(t+1)^T\varphi(t+1)P(t)[1 + \varphi(t+1)P(t)\varphi(t+1)^T]^{-1} \qquad (7.3.8)$$

Applying this result we get

$$K(t+1) = P(t+1)\varphi(t+1)^T$$

$$= P(t)\varphi(t+1)^T[1 + \varphi(t+1)P(t)\varphi(t+1)^T]^{-1} \qquad (7.3.9)$$

Now we obtain the second version of the recursive LS method

$$\hat{\theta}(t+1) = \hat{\theta}(t) + K(t+1)\epsilon(t+1) \tag{7.3.10a}$$

$$\epsilon(t+1) = y(t+1) - \varphi(t+1)\hat{\theta}(t) \tag{7.3.10b}$$

$$K(t+1) = P(t+1)\varphi(t+1)^T = P(t)\varphi(t+1)^T[1 + \varphi(t+1)P(t)\varphi(t+1)^T]^{-1} \tag{7.3.10c}$$

$$P(t+1) = P(t) - P(t)\varphi(t+1)^T\varphi(t+1)P(t)[1 + \varphi(t+1)P(t)\varphi(t+1)^T]^{-1} \tag{7.3.10d}$$

Here no matrix inversion is needed.

We note that in the derivation of the recursive LS algorithm, no approximation has been made. Therefore, the recursive LS estimate and the off-line estimate are theoretically identical. This is another advantage of the LS method.

The updating (7.3.10d) for $P(t+1)$ is not always numerically robust. Rounding errors may accumulate and make the computed $P(t+1)$ indefinite, even though it is theoretically always positive definite. When $P(t+1)$ becomes indefinite the parameter estimates tend to diverge. A way to overcome this difficulty is to use a square root algorithm. Define $S(t+1)$ through

$$P(t+1) = S(t+1)S(t+1)^T \tag{7.3.11}$$

and update $S(t+1)$ instead of $P(t+1)$. For details of square root algorithms we refer the interested reader to the book of Ljung and Söderström (1983).

The initial values $P(0)$ and $\hat{\theta}(0)$ can be obtained by the off-line LS method, or by simply taking

$$\hat{\theta}(0) = 0, \quad P(0) = \rho I \tag{7.3.12}$$

where ρ is a "large" number.

Forgetting factor

When the process is slowly time varying, the measurements obtained a long time ago contain less information about the process than the recent measurements. In order to let the estimator follow the change of the process, it is desirable to discount the old measurements in the estimation algorithm. This can be realised by introducing a forgetting factor λ and modifying the LS loss function as follows

$$V_{LS}(t) = \frac{1}{t}\sum_{s=1}^{t}\lambda^{t-s}\epsilon(s)^2 \tag{7.3.13}$$

Here λ is a number just less than one (for example 0.99). This means that with increasing t the measurements obtained previously are discounted. The recursive LS method with a forgetting factor is the same as in (7.3.10) except that $P(t+1)$ is modified to

$$P(t+1) = \{P(t) - P(t)\varphi(t+1)^T\varphi(t+1)P(t)[\lambda + \varphi(t+1)P(t)\varphi(t+1)^T]^{-1}\}/\lambda \qquad (7.3.14)$$

7.3.2 A Recursive Output Error Method

In Section 5.2 we have shown that the Gauss-Newton algorithm for the (off-line) output error method is

$$\hat{\theta}^{k+1} = \hat{\theta}^k - \left[\sum_{t=1}^{N} [\varphi^k(t)]^T\varphi^k(t)\right]^{-1} \sum_{t=1}^{N} [\varphi^k(t)]^T\varepsilon(t,\hat{\theta}^k) \qquad (7.3.15)$$

where $\varepsilon(t,\hat{\theta}^k)$ is the output error residual

$$\varepsilon(t,\hat{\theta}^k) = y(t) - \hat{G}^k(q)u(t) \qquad (7.3.16)$$

$\varphi^k(t)$ is the gradient of the output error residual

$$\varphi^k(t) = \left[\frac{\hat{B}^k(q)}{\hat{A}^k(q)^2}u(t-1) \cdots \frac{\hat{B}^k(q)}{\hat{A}^k(q)^2}u(t-n) \frac{-1}{\hat{A}^k(q)}u(t-1) \cdots \frac{-1}{\hat{A}^k(q)}u(t-n)\right] \qquad (7.3.17)$$

At time $t+1$, we can approximate (7.3.15) by replacing $\hat{\theta}^k$ by $\hat{\theta}(t)$, $\hat{\theta}^{k+1}$ by $\hat{\theta}(t+1)$ and $-\varphi^k(t+1)$ by

$$\varphi(t+1) = \left[\frac{-\hat{B}_t(q)}{\hat{A}_t(q)^2}u(t) \cdots \frac{-\hat{B}_t(q)}{\hat{A}_t(q)^2}u(t-n+1) \frac{1}{\hat{A}_t(q)}u(t) \cdots \frac{1}{\hat{A}_t(q)}u(t-n+1)\right] \qquad (7.3.18)$$

This yields

$$\hat{\theta}(t+1) = \hat{\theta}(t) + \left[\sum_{k=1}^{t+1} \varphi(k)^T\varphi(k)\right]^{-1} \sum_{k=1}^{t+1} \varphi(k)^T\varepsilon_{oe}(s,\hat{\theta}(t))$$

Introducing

$$P(t+1) = \left[\sum_{k=1}^{t+1} \varphi(k)^T\varphi(k)\right]^{-1} \qquad (7.3.19)$$

we have

$$\hat{\theta}(t+1) = \hat{\theta}(t) + P(t+1)\left[\sum_{k=1}^{t}\varphi(k)^{T}\varepsilon_{oe}(k,\hat{\theta}(t)) + \varphi(t+1)^{T}\varepsilon_{oe}(t+1,\hat{\theta}(t))\right] \tag{7.3.20}$$

Observe that the off-line estimate $\hat{\theta}(t)$ is obtained such that

$$\frac{\partial}{\partial\theta}\left[\sum_{k=1}^{t}\varepsilon_{oe}(k,\hat{\theta}(t))^{2}\right] = 2\sum_{k=1}^{t}\varphi(k)^{T}\varepsilon_{oe}(k,\hat{\theta}(t)) = 0$$

Using this equation in (7.3.20) with some approximation, we obtain

$$\hat{\theta}(t+1) = \hat{\theta}(t) + P(t+1)\varphi(t+1)^{T}\varepsilon_{oe}(t+1,\hat{\theta}(t)) \tag{7.3.21}$$

This equation has the same algebraic structure as (7.3.7a). Therefore, we obtain immediately the recursive version for the output error method (7.3.15)

$$\hat{\theta}(t+1) = \hat{\theta} + K(t+1)\varepsilon_{oe}(t+1,\hat{\theta}(t)) \tag{7.3.22a}$$

$$\varepsilon(t+1,\theta(t)) = y(t+1) - \hat{G}_{f}(q)u(t+1) \tag{7.3.22b}$$

$$K(t+1) = P(t+1)\varphi(t+1)^{T} = P(t)\varphi(t+1)^{T}[1 + \varphi(t+1)P(t)\varphi(t+1)^{T}]^{-1} \tag{7.3.22c}$$

$$P(t+1) = P(t) - P(t)\varphi(t+1)^{T}\varphi(t+1)P(t)[1 + \varphi(t+1)P(t)\varphi(t+1)^{T}]^{-1} \tag{7.3.22d}$$

where $\varphi(t)$ is given in (7.3.18).

Similarly as for the recursive LS method, we can use a forgetting factor to track a time-varying process.

7.3.3 A Recursive Identifier for Robust Adaptive Control

After having studied the recursive LS method and the recursive output error method, it is not difficult to implement our identification method in a recursive form. Assume that the process is operating in a closed loop and that the test signal $w(t)$ is generated by

$$w(t) = F_{w}(q)\xi(t) \tag{7.3.23}$$

where $\xi(t)$ is a white noise and $F_{w}(q)$ is the given filter for realizing a desired spectrum.

The initial estimate. The initial estimate of the model can be obtained by the off-line algorithm of Section 2, using N_{0} samples.

Then, starting at time N_0+1, the model can be updated as follows:

Step 1. Estimation of a high order model

Previously we have seen that the off-line algorithm of LS (equation error) method can be written in a recursive form without approximation. The estimate $\theta^n(t+1)$ is determined by (7.3.10).

Step 2. Model reduction

Our model reduction method in Section 7.2 is an output error estimation. Hence it can be implemented in a recursive fashion as follows.

1   Simulation. For $k = mt+1, \cdots, mt+m$, calculate the input as

$$u_f(k) = [\hat{A}^n_{t+1}(q)\hat{S}_t(q)F_w(q)]\xi(k) \qquad (7.3.24)$$

where $\hat{A}^n_{t+1}(q)$ is the denominator polynomial of the high order model at time $t+1$, $\hat{S}_t(q)$ is the sensitivity function evaluated using the high order model at time t, $F_w(q)$ is the filter for generating the test signal $w(t)$, and $\xi(k)$ is a white noise sequence. Then simulate the high order model

$$\hat{y}^n(k) = \hat{G}^n_{t+1}(q)u_f(k) = [\hat{B}^n_{t+1}(q)\hat{S}_t(t)F_w(q)]\xi(k) \qquad (7.3.25)$$

2   Recursive output error model estimation. Denote the parameter vector of the reduced model as

$$\theta^l = [a^l_1 \cdots a^l_{nl} \; b^l_1 \cdots b^l_{nl}]^T \qquad (7.3.26)$$

where m is the order of the reduced model, denote the output error residual as

$$\epsilon_{OE}(k) = \hat{y}^n(s) - \frac{\hat{B}^l_t(q)}{\hat{A}^l_t(q)} u_f(k) \qquad (7.3.27)$$

and denote the data vector of the reduced model as

$$\varphi_f(k) = [\frac{\partial \epsilon_{oe}}{\partial \theta^l}(k,\theta^l(k))]^T$$

$$= [\frac{-\hat{B}_k^l(q)}{[\hat{A}_m^l(q)]^2}u_r(k-1)\ \cdots\frac{-\hat{B}_k^l(q)}{[\hat{A}_m^l(q)]^2}u_r(k-m)\ \frac{1}{\hat{A}_m^l(q)}u_r(k-1)\ \cdots\frac{1}{\hat{A}_m^l(q)}u_r(k-m)]$$

$$(7.3.28)$$

Then a recursive output error method based on the Gauss-Newton algorithm for $k = mt+1$ to $k = mt+m$ is given by (7.3.22).

Step 3. Calculating the upper bound

Based on the formula of upper bound (7.2.8) we have

$$\bar{\Delta}_{t+1}(\omega) = 3\sqrt{\frac{n}{t+1}\frac{\bar{\Phi}_v(\omega,t+1)}{\hat{\Phi}_w(\omega)\,|\hat{S}_{t+1}(e^{i\omega})|^2}}$$

$$= 3\sqrt{\frac{n\bar{R}_\varepsilon}{(t+1)R_\xi}}\frac{1}{|\hat{A}_{t+1}^n(e^{i\omega})|\,|F_w(e^{i\omega})|\,|\hat{S}_{t+1}(e^{i\omega})|} \qquad (7.3.29)$$

where $\hat{R}_\varepsilon$ is estimated variance of equation error residual of the high order model, R_ξ is the variance of the white noise used to generate the test signal $w(t)$, and $\hat{S}_{t+1}(e^{i\omega})$ is determined by

$$\hat{S}_{t+1}(q) = \frac{1}{1 + \hat{G}_{t+1}^l(q)C_{t+1}(q)}$$

Remark— In this algorithm, at each sample time, m samples are simulated from the high order model, and m iterations are used to update the reduced model. Theoretically speaking, a large m is required to approximate the loss function (7.2.6). The speed of the computer will set a limit on m. How to make a good choice for m needs to be studied further. From our experience in simulation studies, $m = 5$ is a reasonable choice.

7.4 A Simulation Study

The process to be identified is

$$G_o(q) = \frac{q^{-1} + 0.5q^{-2}}{1 - 1.5q^{-1} + 0.7q^{-2}} \qquad (7.4.1)$$

The output disturbance is

$$v(t) = \frac{0.5q^{-1} + q^{-2}}{1 - 1.2q^{-1} + 0.36q^{-2}} \xi_1(t) \qquad (7.4.2)$$

and the test input $w(t)$ is

$$w(t) = \frac{1 + 0.4q^{-1} + 0.3q^{-2}}{1 - 1.4q^{-1} + 0.49q^{-2}} \xi_2(t) \qquad (7.4.3)$$

where $\{\xi_1(t)\}$ and $\{\xi_2(t)\}$ are two independent zero mean white Gaussian noise sequences.

The noise-to-signal ratio at the output is always 10% (in power):

$$N/S = \text{var}(v(t))/\text{var}(y_o(t)) = 10\%$$

With coloured disturbance, this can be considered as very noisy measurement.

The amplitudes of the frequency responses of the process, test signal filter and disturbance filter are shown in Fig. 7.4.1

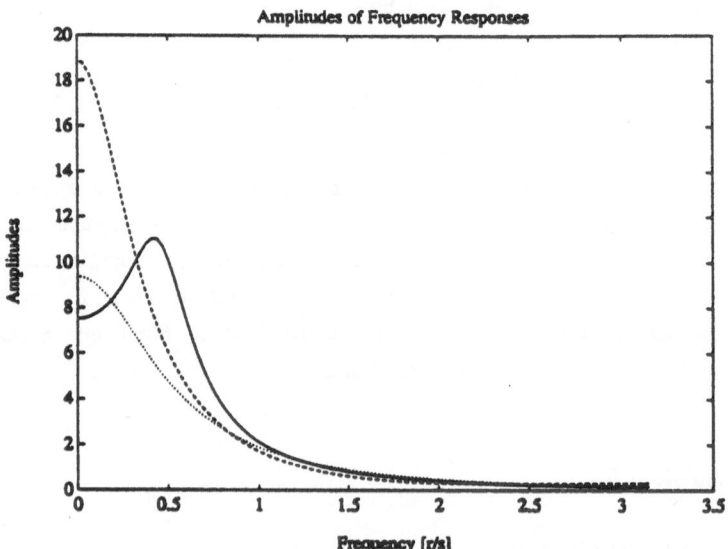

Fig. 7.4.1 The amplitudes of the frequency responses of the process (solid), the signal filter (dashed), and the disturbance filter (dotted).

1) Testing the New Method of Recursive Estimation

The input/output data is generated under a proportional feedback $C = 0.2$. The order of high order model is taken as 15; the Steiglitz-McBride method is used in model reduction, where the number of iterations for each sample is $m = 5$. The first 80 samples are used to get the initial estimates using the off-line algorithm of Section 7.2.

Our method is then validated in the frequency domain. The errors of both the high order model and the reduced model (2nd order) are compared with the estimated upper bound for $N = 100, 200, 500$, and 1500; see Fig. 7.4.2. We find that the errors of the two models decrease when the number of data increases, which is reflected by the upper bound. We also see that the reduced model has a higher accuracy.

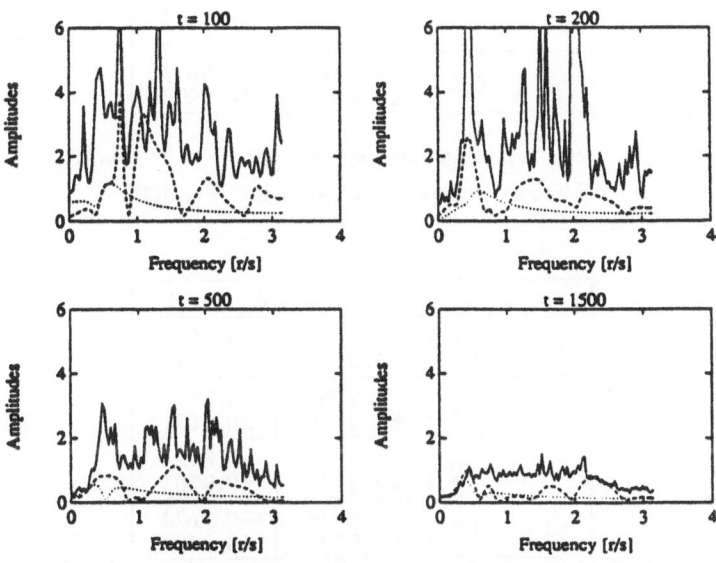

Fig. 7.4.2   The upper bound (solid), error of high order model (dashed) and the error of the reduced model (dotted); $N = 100, 200, 500$ and 1500; closed-loop experiment.

2) Comparing the Off-Line Algorithm with Prediction Error Method

The off-line algorithm in Section 7.2 is compared with the off-line prediction error method implemented in Matlab Identification Tool Box. A

Box-Jenkins model with correct model orders for the process and the disturbance filter is used. Both open-loop and closed-loop (C = 0.2) situations are simulated with the number of samples N = 1000. The order of the high order model are taken as 10, 15, and 20. Order 20 gives the best result; hence this order will be used for the comparison. For each case, 20 simulation runs are used for the comparison. The mean values and standard deviations of parameter estimates of the three methods are shown in Table 7.4.1 and Table 7.4.2. The model qualities of the two methods are very similar.

True Value	New method	B-J method
a1 = -1.5	-1.4987 ±0.0293	-1.5021 ±0.0258
a2 = 0.7	0.6990 ±0.0195	0.7017 ±0.0180
b1 = 1.0	0.9950 ±0.1329	-0.9977 ±0.1192
b2 = 0.5	0.5147 ±0.2212	0.5012 ±0.1972

Table 7.4.1   Mean values and standard deviations of 20 simulation runs, off-line estimators, open loop data, N = 1000.

True Value	New method	B-J method
a1 = -1.5	-1.5006 ±0.0140	-1.5035 ±0.0112
a2 = 0.7	0.7005 ±0.0116	0.7017 ±0.0091
b1 = 1.0	0.8558 ±0.0675	-0.8722 ±0.0505
b2 = 0.5	0.6048 ±0.0839	-0.5628 ±0.0600

Table 7.4.2   Mean values and standard deviations of parameters of 20 simulation runs, off-line algorithms, closed-loop data, N = 1000.

3)   Comparing the Recursive Algorithm with Prediction Error Method

In the test of the recursive algorithm, it turns out that when the order of the high order model is 20, we occasionally encounter outliers in the recursive model reduction step, i.e., the reduced model is far from the true process. The outlier problem is less severe when the high order is decreased. While leaving this problem for future research, we will use order 15 for the high order model in the recursive algorithm; remember that this order is too low for the best accuracy, according to the simulation test of the off-line algorithm. The new recursive algorithm will be compared with the *recursive Box-Jenkins method* implemented in the Matlab Identification Toolbox. For all the algorithms, their corresponding off-line algorithms are used to calculate the initial parameters using the first 80 samples; and then the initial parameters and the 80 samples are used to calculate the initial P matrices. Closed loop simulations ($C = 0.2$) are performed and the number of samples is $N = 200$, 500, and 1500. For each case 20 runs are used for the comparison. The results are shown in Table 7.4.3, 7.4.4 and 7.4.5.

True Value	New method	B-J method
a1 = -1.5	-1.5131 ±0.0574	-1.4834 ±0.0504
a2 = 0.7	0.7357 ±0.0905	0.6823 ±0.0463
b1 = 1.0	0.8071 ±0.3155	0.8972 ±0.2518
b2 = 0.5	0.5001 ±0.2752	0.5220 ±0.3346

Table 7.4.3   Mean values and standard deviations of 20 simulation runs, recursive estimators, closed-loop data, $N = 200$.

True Value	New method	B-J method
a1 = -1.5	-1.5038 ±0.0580	-1.4994 ±0.0455
a2 = 0.7	0.7334 ±0.0760	0.6980 ±0.0317
b1 = 1.0	0.7644 ±0.3187	0.8952 ±0.1602
b2 = 0.5	0.5426 ±0.2409	0.5398 ±0.2421

Table 7.4.4   Mean values and standard deviations of 20 simulation runs, recursive estimators, closed-loop data, $N = 500$.

134

True Value	New method	B-J method
a1 = -1.5	-1.4940 ±0.0301	-1.4991 ±0.0200
a2 = 0.7	0.7399 ±0.0886	0.6989 ±0.0133
b1 = 1.0	0.6992 ±0.2978	0.8507 ±0.1281
b2 = 0.5	0.5760 ±0.2004	0.5929 ±0.1724

Table 7.4.5 Mean values and standard deviations of 20 simulation runs, recursive estimators, closed-loop data, $N = 1500$.

We find that the recursive Box-Jenkins method performs better than our method, but the difference is not significant. The computing time of the new method is about 3 times longer than that of the Box-Jenkins method. Fig. 7.4.3 shows the parameter estimates of the two methods from one closed loop simulation.

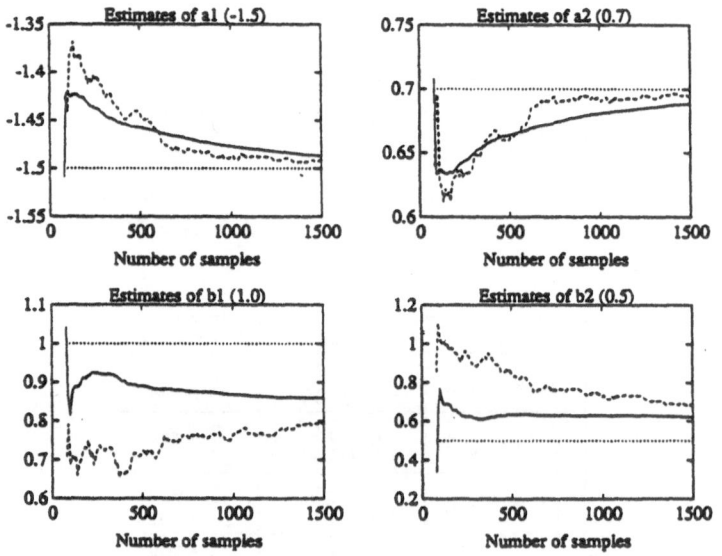

Fig. 7.4.3 The behaviour of the two recursive estimators; "solid line" := the new method, "dashed line" := the Box-Jenkins method; closed-loop experiment.

Comments on the simulation results:

— The off-line algorithm of our method has reached the accuracy of a prediction error method in this example. This implies that our method can not only supply an error bound, but has high model accuracy.

— The outlier problem of the recursive algorithm takes place only at the step of model reduction, because the recursive high order model estimate is identical to the off-line estimate which reaches the global minimum. A better way of implementing the step of model reduction need to be researched in order to overcome the outlier problem.

— The outlier problem of the reduced model can easily be detected by the same method for detecting a local minimum which was proposed in Section 7.2.

7.5 Conclusions

In this chapter we have developed a method of identification which is suitable for use in robust control. Let us call it a *two-step method*. The method is based on the asymptotic theory of Ljung (1985). Fundamental problems, such as input design, model order selection, parameter estimation and model uncertainty, can be solved by the method. In all the steps we use criteria which are intuitively sensible; and the computations needed are simple and reliable. The simulation study shows that the off-line algorithm of the method can reach the accuracy of the prediction error methods. The method can also be used for the identification of infinite dimensional processes, because the asymptotic theory is not restricted to finite dimensional processes.

After deriving the recursive LS method and the recursive output error method, we have proposed a recursive algorithm for the two-step method. The numerical quality of the new recursive algorithm needs to be improved. This is now under investigation. Another topic under study is the extension of the method to include a forgetting factor for a time-varying situation. An (potential) additional advantage of the new approach for time-varying processes is that we have more freedom in making a trade-off between tracking ability and variance reduction.

It is quite straightforward to extend the method to MIMO case. This will be done in the next chapter.

CHAPTER 8

IDENTIFICATION FOR ROBUST CONTROL; MIMO CASE

In this chapter, we will solve the problem of multi-input multi-output (MIMO) process identification for robust control. We will show that all the steps of the *two-step method* developed in the previous chapter can be generalized for MIMO processes. For the sake of simplicity, we will restrict ourselves to open-loop experiments. In Section 8.1 the MIMO version of the asymptotic theory is presented. The identification procedure is presented in Section 8.2. In Section 8.3, two MIMO industrial processes will be identified using our method, and the results will be compared with those of the prediction methods. In Section 8.4 we will show that closed-loop identification can be transformed into two open-loop problems. Section 8.5 gives the conclusions.

8.1 The MIMO Version of the Asymptotic Theory

Consider a linear time-invariant discrete-time process with m inputs and p outputs

$$y(t) = G^{\circ}(q)u(t) + v(t) \qquad (8.1.1)$$

where

$$G^{\circ}(q) = \sum_{k=1}^{\infty} G_k^{\circ} \cdot q^{-k}$$

is called the transfer operator of the process ; $y(t)$ is the p-dimensional column output vector at time t; $u(t)$ is the m-dimensional column input vector at time t; $\{G_k^{\circ}\}$ is the impulse response of the process, which is a sequence of $p \times m$ matrices; $\{v(t)\}$ is a p-dimensional stochastic stationary process with zero mean values.

The frequency response matrix of the process is defined as

$$G^{\circ}(e^{i\omega}) = \sum_{k=1}^{\infty} G_k^{\circ} \cdot e^{-i\omega k} \qquad -\pi \leq \omega \leq \pi \qquad (8.1.2)$$

The disturbances vector $v(t)$ comprises filtered white noise signals

$$v(t) = H^{\circ}(q)\xi(t) \qquad (8.1.3)$$

where $\xi(t)$ is a p-dimensional white noise vector and $H^{\circ}(q)$ is the disturbance filter matrix which is stable and minimum phase (its inverse is
stable).

If we have a parametrized model:

$$y(t) = G(q,\theta)u(t) + H(q,\theta)e(t) \qquad (8.1.4)$$

and denote the data sequence as

$$Z^N := y(1), \ u(1), \ \dots \dots, \ y(N), \ u(N)$$

then we can determine the parameters by minimizing the sum of the squared
prediction errors

$$V^N(\theta) = \frac{1}{N} \sum_{t=1}^{N} \varepsilon^T(t,\theta)\varepsilon(t,\theta) \qquad (8.1.5)$$

where $\varepsilon(t,\theta)$ is the prediction error vector:

$$\varepsilon(t,\theta) = y(t) - \hat{y}(t\,|\,\theta) = H^{-1}(q,\theta)[y(t) - G(q,\theta)u(t)] \qquad (8.1.6)$$

After the parameter estimation, the transfer function estimates are denoted
as:

$$\hat{G}^n_N(e^{i\omega}) = G(\hat{\theta},e^{i\omega}), \qquad \hat{H}^n_N(e^{i\omega}) = H(\hat{\theta},e^{i\omega}) \qquad (8.1.7)$$

The general MIMO model is defined as

$$F(q)y(t) = A^{-1}(q)B(q)u(t) + D^{-1}(q)C(q)e(t) \qquad (8.1.8)$$

where $A(q)$, $B(q)$, $C(q)$, $D(q)$ and $F(q)$ are polynomial matrices with dimension
$p{\times}p$, $p{\times}m$, $p{\times}p$, $p{\times}p$ and $p{\times}p$ respectively. The parameter vector θ is formed by
the coefficients of the polynomial matrices. This form is too general for
real applications. In the two-step method developed in Chapter 7, the final
model has a Box-Jenkins structure. The MIMO Box-Jenkins model is given as

$$y(t) = A^{-1}(q)B(q)u(t) + D^{-1}(q)C(q)e(t) \qquad (8.1.9)$$

We need further specifications for matrices $A(q)$, $B(q)$, $C(q)$ and $D(q)$ in
order to have a MIMO model which is identifiable, meaning roughly that the
model is uniquely determined if a prediction error method is used and if the
data are persistently exciting. For example, we can let $A(q)$ and $D(q)$ be
triangular or diagonal polynomial matrices, and the degrees of the polynomial entries follow some rules. This leads to different *canonical forms* of
MIMO models which are in general rather complex. In our method, we will

avoid the study of these canonical forms and use one of the simplest form: the diagonal form (this is also a canonical form). A diagonal form of the MIMO Box-Jenkins model is that $A(q)$, $C(q)$ and $D(q)$ are diagonal polynomial matrices, and their nonzero polynomials are all monic (the highest degree term is 1).

Asymptotic Properties of the Transfer Function Estimates

For the presentation of the asymptotic theory, we do not restrict ourselves to the diagonal form. So the following result holds for more general models. Assume that all the non-zero polynomials of the matrices of the MIMO model have the same degree n. Note that, unlike the SISO case, this *degree* is in general not the same as the McMillan degree, or the order which is defined as the dimension of the minimal state space realization of the model. Generally, the McMillan degree of the model equals $p \cdot n$.

Result 8.1.1 Given (8.1.1) - (8.1.8). Assume that degree n follows

$$n \to \infty \quad \text{as } N \to \infty \quad \text{and} \quad n^2/N \to 0 \quad \text{as } N \to \infty$$

and that the input is persistently exciting with any finite order, so that,

$$\sigma_{min}[\Phi_u(\omega)] > 0 \quad \text{for} \quad -\pi < \omega < \pi$$

where $\sigma_{min}(\cdot)$ denotes the smallest singular value of the given matrix. And also assume that the global minima are obtained for all n and N. Then

$$[\hat{G}_N^n(e^{i\omega}), \hat{H}_N^n(e^{i\omega})] \to [G^o(e^{i\omega}), H^o(e^{i\omega})] \quad w.p.1 \text{ as } N \to \infty \qquad (8.1.10)$$

(consistency)

$$col \ \sqrt{N}\left[\hat{G}_N^n(e^{i\omega}), \hat{H}_N^n(e^{i\omega})\right] \to N(0, P_n(\omega)) \quad \text{as } N \to \infty \qquad (8.1.11)$$

(Asymptotic Gaussian distribution)

and

$$\lim_{n \to \infty} \frac{1}{n} P_n(\omega) = [\Phi(\omega)]^T \otimes \Phi_v(\omega) \qquad (8.1.12)$$

(Asymptotic covariance)

where

$$\Phi(\omega) = \begin{bmatrix} \Phi_u(\omega) & \Phi_{u\xi}(\omega) \\ \Phi_{\xi u}(\omega) & R \end{bmatrix}, \quad \Phi_v(\omega) = H^\circ(e^{i\omega})RH^\circ(e^{-i\omega})$$

and R is the covariance of the white noise $\xi(t)$, $col(\cdot)$ denotes the vector operator on a matrix, -T means inverse and transpose, $\otimes$ denotes the Kronecker product. Let

$$A = \{ a_{ij} \}, \quad B = \{ b_{ij} \}$$

be $m \times n$ and $p \times r$ matrices, respectively. The Kronecker product of A and B is defined as an $mp \times nr$ matrix, denoted by $A \otimes B$, so that,

$$A \otimes B = \begin{bmatrix} a_{11}B & \cdots & a_{1n}B \\ \vdots & \ddots & \vdots \\ a_{m1}B & \cdots & a_{mn}B \end{bmatrix}$$

Result 8.1.1 is the MIMO extension of Ljung (1985), the proof can be found in Zhu (1989). For an open-loop experiment we have

$$col[\hat{G}_N^n(e^{i\omega})] \rightarrow N\left(0, \frac{n}{N}[\Phi_u(\omega)]^{-T} \otimes \Phi_v(\omega)\right) \quad \text{as } N \rightarrow \infty \qquad (8.1.13)$$

$$col[\hat{H}_N^n(e^{i\omega})] \rightarrow N\left(0, \frac{n}{N}[R^{-1} \otimes \Phi_v(\omega)]\right) \quad \text{as } N \rightarrow \infty \qquad (8.1.14)$$

8.2 The Identification Method

We shall present the MIMO version of our method only for open-loop experiments. There are two reasons for this. The first reason is to keep the presentation simple. The second reason is that if a closed loop experiment is necessary, we can transform the problem of closed loop identification into two open-loop problems; and this will be shown in Section 8.4. Therefore, it is sufficient to treat only the open-loop problem.

8.2.1 The Procedure

Like the SISO case, the procedure consists of: Step 1 — Estimation of a high order model; Step 2 — Model Reduction and Step 3 — Deriving an upper bound matrix

Step 1. Estimation of a high order model

Given the input/output data from an open-loop experiment

$$Z^N := y(1), \ u(1), \ ... \ ..., \ y(N), \ u(N)$$

Assume that different inputs are mutually independent, which is realistic for open-loop experiment. Estimate the parameters of an equation error model with a high order, $n = 30 \sim 50$. We will, in our method, always use the diagonal form. This means that matrix $A(q)$ is diagonal and the MIMO model is decoupled into p MISO models. After the estimation, we have

$$\hat{G}_N^n(q) = \hat{A}_N^{-1}(q)\hat{B}_N(q) \qquad \hat{H}_N^n(q) = \hat{A}_N^{-1}(q) \tag{8.2.1}$$

and the spectrum estimates of the disturbances are

$$\hat{\Phi}_v(\omega) = \hat{A}_N^{-1}(e^{i\omega})\hat{R} \ \hat{A}_N^{-T}(e^{-i\omega}) \tag{8.2.2}$$

where $\hat{R}$ is the estimated covariance of the equation error residuals.

Step 2. Model reduction

The asymptotic maximum likelihood model reduction can be extended easily to diagonal MIMO models. Denote $\hat{G}_{ij}^n(e^{i\omega})$ and $G_{ij}^o(e^{i\omega})$ as the (i, j) entries of $\hat{G}_N^n(e^{i\omega})$ and $G^o(e^{i\omega})$ respectively. According to the asymptotic theory of Section 8.1, we know that for an open-loop experiment with mutually independent inputs, the high order estimates of each transfer functions are mutually independent, and each transfer function follows a Gaussian distribution with a variance

$$\text{var}[\hat{G}_{ij}^n(e^{i\omega})] \approx \frac{n}{N} \frac{\Phi_{v_i}(\omega)}{\Phi_{u_j}(\omega)} \tag{8.2.3}$$

where $\Phi_{u_j}(\omega)$ is the spectrum of $u_j(t)$, and $\Phi_{v_i}(\omega)$ is the spectrum of $v_i(t)$.

For simplicity we also use the diagonal form for the reduced model. This implies that we can perform model reduction for each MISO model separately. It is not difficult to see that the negative log-likelihood function for the i-th sub-model (related to $y_i(t)$) is

$$V = \sum_{j=1}^{m} \frac{1}{2\pi} \int_{-\pi}^{\pi} |\hat{G}_{ij}^n(e^{i\omega}) - \hat{G}_{ij}^l(e^{i\omega})|^2 \frac{\Phi_{u_j}(\omega)}{|H_{ii}^o(e^{i\omega})|^2} \, d\omega \tag{8.2.4}$$

Again, we use the following identification method for model reduction.

For the i-th MISO sub-model:

1) **Simulation** Collect the inputs, $\{u(t),\ t = 1,\ \cdots,\ N\}$, which have been used in the identification experiment. Filter the inputs by the inverse of the disturbance model $1/\hat{H}_{ii}^n(q) = \hat{A}_{ii}^n(q)$. Then simulate the i-th MISO high order sub-model using the filtered inputs:

$$\hat{y}_i(t) = \frac{1}{\hat{A}_{ii}^n(q)}[\hat{B}_{11}^n(q)\ \cdots\ \hat{B}_{1m}^n(q)][\hat{A}_{ii}^n(q)u(t)] \qquad (8.2.5)$$

This is equivalent to

$$\hat{y}_i(t) = [\hat{B}_{11}^n(q)\cdots\ \hat{B}_{1m}^n(q)]u(t) \qquad (8.2.6)$$

which is simpler than (8.2.5). Thus we obtain the input/output data of the i-th sub-model:

$$Z_i^N = \hat{y}_i(1),\ u_{f,1}(1),\ \cdots,\ u_{f,m}(1),\ \cdots,\ \cdots,\ \hat{y}(N),\ u_{f,1}(N),\cdots\ u_{f,m}(N)$$

where

$$u_f(t) = \hat{A}_{ii}^n(q)u(t)$$

2) **Parameter estimation** The parameters of the reduced model are calculated by using an *output error method* which minimizes the following loss function

$$V_{oe}^N = \frac{1}{N}\sum_{t=1}^{N}\left\{\sum_{j=1}^{m}[\hat{G}_{ij}^n(q) - \hat{G}_{ij}^l(q)]u_{f,j}(t)\right\}^2 \qquad (8.2.7)$$

Letting $N \to \infty$, applying Parseval's identity and remembering that the inputs are mutually independent, one can show that the output error loss function (8.2.7) tends to

$$V_{oe}^\infty = \sum_{j=1}^{m}\frac{1}{2\pi}\int_{-\pi}^{\pi}|\hat{G}_{ij}^n(e^{i\omega}) - \hat{G}_{ij}^l(e^{i\omega})|^2\frac{\Phi_{u_j}(\omega)}{|\hat{H}_{ii}^n(e^{i\omega})|^2}\,d\omega \qquad (8.2.8)$$

This is the log-likelihood function of (8.2.4) for the i-th MISO sub-model, except that $H_{ii}^o(e^{i\omega})$ has been replaced by the high order estimate

$$\hat{H}^n_{ii}(e^{i\omega}) = 1/\hat{A}^n_{ii}(e^{i\omega}).$$

In a similar way, model reduction for the disturbance model can be performed.

Step 3. Deriving an upper bound matrix

According to the asymptotic theory and (8.2.3), we can define the 3σ bound for $\{G^o_{ij}(e^{i\omega}) - \hat{G}^n_{ij}(e^{i\omega})\}$ as

$$|G^o_{ij}(e^{i\omega}) - \hat{G}^n_{ij}(e^{i\omega})| \le 3\sqrt{\frac{n}{N}[\Phi_{u_j}(\omega)]^{-1}\Phi_{v_i}(\omega)} \quad \text{w.p. } 99.99\% \qquad (8.2.9)$$

Because our model reduction will improve the model accuracy, this bound can be used for the reduced model. Define the error matrix of the reduced model as

$$\Delta(e^{i\omega}) = G^o(e^{i\omega}) - \hat{G}^l(e^{i\omega}) \qquad (8.2.10)$$

Then we have an upper bound matrix

$$\overline{\Delta}(\omega) = \{\overline{\Delta}_{ij}(\omega)\} \qquad (8.2.11)$$

where

$$\overline{\Delta}_{ij}(\omega) = 3\sqrt{\frac{n}{N}[\Phi_{u_j}(\omega)]^{-1}\Phi_{v_i}(\omega)} \qquad (8.2.12)$$

such that

$$|\Delta_{ij}(e^{i\omega})| \le \overline{\Delta}_{ij}(\omega) \quad \text{for all } i, j \quad \text{w.p. } 99.99\% \qquad (8.2.13)$$

8.2.2 Model Structure Selection

As mentioned before, we will use the diagonal form for the reduced model. Denote l_i as the order (the degree of the polynomials) of the i-th sub-model, then $[l_1, \cdots l_p]$ determines the structure of the diagonal form MIMO model. The following rule is a natural extension of the order selection rule for SISO processes.

A MIMO model structure selection rule

For the i-th sub-model: Perform model reduction with various orders and select the order, l_i, such that in the frequency range which is important for control system design, the following relation holds:

$$\sum_{j=1}^{m} |\hat{G}^{n}_{ij}(e^{i\omega}) - {}^{TM}\hat{G}^{l}_{ij}(e^{i\omega})|^2 \approx \frac{n}{N} \sum_{j=1}^{m} [\hat{\Phi}_{u_j}(\omega)]^{-1}\hat{\Phi}_{v_i}(\omega) \qquad (8.2.14)$$

The order of the disturbance model can simply be taken as l_i, or, selected in a similar way as in (8.2.14).

The reduced model is given in polynomial matrix description in a diagonal form. In the development of the identification method, we have usefully exploited the simplicity of this model structure. One possible drawback of this model structure is that its minimality is not always guaranteed, which means that the sum of the degrees of the sub-models is greater than the McMillan degree of the model. This problem can easily be solved by transforming the model to a state space model and then performing a model reduction on the state space model. This will be discussed in Section 8.2.4.

8.2.3 Input Design for Simulation

We know that a model which can optimally simulate the underlying process for certain given inputs, will be very suitable for an internal model control scheme; see Fig. 7.2.1. Denote $u^s(t)$ as the control input to the process with spectrum $\Phi^s_u(\omega)$. It can be shown (see Lenssen, 1988) that for MIMO processes, under the assumption that the inputs are mutually independent, the optimal input spectrum for the i-th input is given by

$$\Phi^{op}_{u_i}(\omega) = \mu \sqrt{\Phi^s_{u_i}(\omega) \cdot \sum_{j=1}^{p} \Phi_{v_j}(\omega)} \qquad (8.2.15)$$

A way to approximate this is simply

$$\Phi^{op}_{u_i}(\omega) \approx \mu \sum_{j=1}^{p} \Phi_{v_j}(\omega) \qquad (8.2.16)$$

8.2.4 Determining a State Space Realization

For MIMO controller design, it is often convenient to have state space realizations of the process model and the disturbance model. Further model reduction on the state space model will result more compact model.

The reduced model is a diagonal form Box-Jenkins model

$$y(t) = [\hat{A}^l(q)]^{-1}\hat{B}^l(q)u(t) + [\hat{D}^l(q)]^{-1}\hat{C}^l(q)\varepsilon(t) \qquad (8.2.17)$$

Here $\hat{A}^l(q)$, $\hat{C}^l(q)$ and $\hat{D}^l(q)$ are diagonal matrices and their polynomials are all monic. The transfer operator of the process model is

$$\hat{G}^l(q) = [\hat{A}^l(q)]^{-1}\hat{B}^l(q)$$

$$= \text{diag}\left[\frac{1}{\hat{A}^l_{11}(q)} \cdots \frac{1}{\hat{A}^l_{pp}(q)}\right] \begin{bmatrix} \hat{B}^l_{11}(q) & \cdots & \hat{B}^l_{1m}(q) \\ \vdots & & \\ \hat{B}^l_{m1}(q) & \cdots & \hat{B}^l_{mm}(q) \end{bmatrix} \qquad (8.2.18)$$

and the transfer operator of the disturbance is

$$\hat{H}^l(q) = \text{diag}\left[\frac{\hat{C}^l_{11}(q)}{\hat{D}^l_{11}(q)} \cdots \frac{\hat{C}^l_{pp}(q)}{\hat{D}^l_{pp}(q)} \right] \qquad (8.2.19)$$

The following steps are proposed to derive a state space model from the process model $\hat{G}^l(q)$.

1) *Writing down a state space model for each MISO model.* Denote the i-th sub-model of the process as

$$\hat{y}_i(t) = [\hat{A}^l_{ii}(q)]^{-1}\hat{B}^l_i(q)u(t) \qquad (8.2.20)$$

where $\hat{A}^l_{ii}(q)$ is a polynomial

$$\hat{A}^l_{ii}(q) = 1 + \hat{a}_{ii,1}q^{-1} + \cdots \hat{a}_{ii,li}q^{-li}$$

$\hat{B}^l_i(q)$ is a 1×m polynomial vector

$$\hat{B}^l_i(q) = \left[\hat{b}_{i1,1} \cdots \hat{b}_{ip,1}\right]q^{-1} + \cdots + \left[\hat{b}_{i1,li} \cdots \hat{b}_{ip,li}\right]q^{-li}$$

$$= \hat{B}_{i,1}q^{-1} + \cdots + \hat{B}_{i,li}q^{-li}$$

Then the observer canonical form of the sub-model is simply

$$\begin{cases} x_i(t+1) = A_i x_i(t) + B_i u(t) \\ \hat{y}_i(t) = C_i x_i(t) \end{cases} \qquad (8.2.21)$$

with

$$A_i = \begin{bmatrix} -\hat{a}_{ii,1} & 1 & 0 & \cdots & 0 \\ -\hat{a}_{ii,2} & 0 & 1 & \cdots & 0 \\ \vdots & \vdots & & \ddots & \\ & & & & 1 \\ -\hat{a}_{ii,li} & 0 & 0 & \cdots & 0 \end{bmatrix}, \quad B_i = \begin{bmatrix} \hat{B}_{i,1} \\ \hat{B}_{i,2} \\ \vdots \\ \hat{B}_{i,li} \end{bmatrix}$$

$$C_i = [\ 1 \quad 0 \ \cdots \ 0\ ]$$

2) *Forming a state space model of the total MIMO model.* This is easily done as follows

$$\begin{cases} x(t+1) = Ax(t) + Bu(t) \\ \hat{y}(t) = Cx(t) \end{cases} \tag{8.2.22}$$

with

$$A = \begin{bmatrix} A_1 & & & \\ & A_2 & & \\ & & \ddots & \\ & & & A_p \end{bmatrix}, \quad B = \begin{bmatrix} B_1 \\ B_2 \\ \vdots \\ B_p \end{bmatrix}, \quad C = \begin{bmatrix} C_1 & & & \\ & C_2 & & \\ & & \ddots & \\ & & & C_p \end{bmatrix}$$

3) *Checking the minimality of the model; and performing a model reduction, if necessary.* The order of the total model (8.2.22) is

$$n_t = n_1 + n_2 + \cdots n_p \tag{8.2.23}$$

Theoretically speaking, this order can be higher than the minimal order (McMillan degree) of the state space realization of the model, because it is not always possible to write a given process in a minimal diagonal form. In practice, the identified diagonal form will be minimal even if the true model cannot be written in a minimal diagonal form. This is because the disturbances cause model errors and model reduction takes place in identifying each MISO model. However, it is sensible to perform a model reduction if the the total model (8.2.22) can be well approximated by a model with lower order.

The model reduction method based on a balanced realization can be used here. In this method, the state space model is transformed to the so called balanced realization. In such a realization, all the states are equally observable and controllable. Calculate the singular values of the Hankel matrix (see Chapter 2) of the total model, and denote them as

$$\sigma_1 \geq \sigma_2 \geq \cdots \geq \sigma_{n_t}$$

If there are several very small Hankel singular values, we can eliminate the states corresponding to these Hankel singular values by simply truncating the balanced state space model. Denote $\hat{G}(e^{i\omega})$ as the transfer function of the reduced model and denote n_l as its order, it is known that the error caused by the balanced model reduction is bounded by (see Glover, 1984)

$$\sup_{\omega} \sigma_{max}[\hat{G}^l(e^{i\omega}) - \hat{G}(e^{i\omega})] \leq 2\left[\sum_{k=n_l+1}^{n_t} \sigma_k\right] \qquad (8.2.24)$$

Using this relation and the upper bound matrix, we can determine the reduced order n_l such that

$$2\left[\sum_{k=n_l+1}^{n_t} \sigma_k\right] \leq \frac{1}{3}\sup_{\omega} \sigma_{max}(\Delta(\omega)) \qquad (8.2.25)$$

Using this rule, we can obtain a more compact model without a sacrifice of model accuracy.

8.3   Identification of Two Industrial Processes

In this section, we will put our identification method to test. First the glass tube process will be identified using the two-step method, and the result will be compared with that of the prediction error methods. Then a four-effect evaporator with three inputs and three outputs will be identified using both our method and the prediction error methods.

8.3.1   Identification of the Glass Tube Manufacturing Process

The glass tube process studied in Section 4.3, Section 5.5 and Section 6.3 will be used again as a test case. We have seen that both the LS method and the ARMAX method with structure [4, 4] cannot deliver a good model, but the Markov parameter approach can generate a much better model with order 8. The same data will be used here; where the first 600 samples are used for model estimation and the remaining 669 samples for model validation.

First a high order ARX model with structure [30, 30] is estimated. Then model reduction is performed; and two reduced models are obtained which have structure [7, 7] and [9, 7]. The estimated models are validated on the validation data; and the relative simulation errors are 15.2% and 10.8.% for the model with structure [7, 7], and 8.0% and 10.8% for the model with structure [9, 7]. Fig. 8.3.1 shows the model validation for the second model.

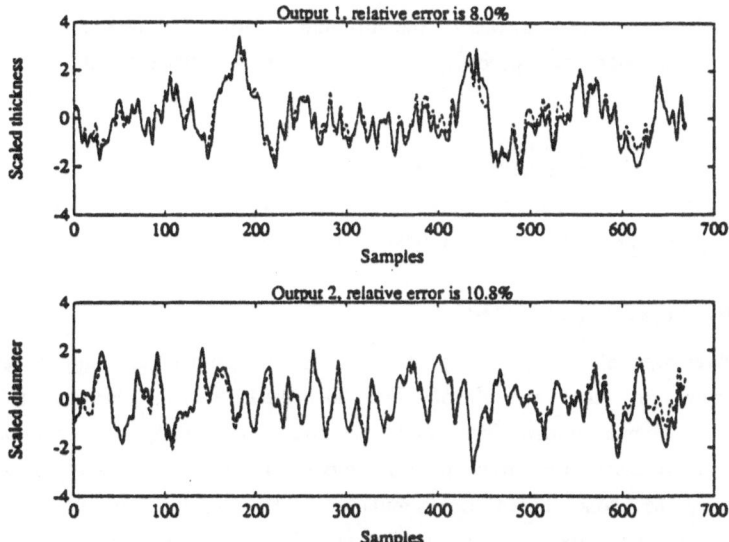

Fig. 8.3.1   Model validation of the two-step method. "solid line" := measured output; "dashed line" := model output.

Now we shall make some comparisons. We recall the ARMAX model of the process:

$$A_{11}(q)y_1(t) = B_{11}(q)u_1(t) + B_{12}(q)u_2(t) + C_{11}(q)\varepsilon_1(t)$$

$$A_{22}(q)y_2(t) = B_{21}(q)u_1(t) + B_{22}(q)u_2(t) + C_{22}(q)\varepsilon_2(t)$$

Two ARMAX models with structure [7, 7] and [9, 9] are estimated. In the model validation, the relative simulation errors are 9.4% and 15.2.% for the model with structure [7, 7], and 9.5% and 15.4% for the model with structure [9, 9].

The following table summarizes the results of different methods which includes also the result obtained by the least-squares method (ARX model) of Chapter 4, and the Markov parameter method of Chapter 6.

Markov order 8	T-S [7,7]	T-S [9,7]	LS [4,4]	ARMAX [7,7]	ARMAX [9,9]
10.1%	15.2%	8.0%	39.8%	9.4%	9.5%
12.6%	10.8%	10.8%	41.7%	15.2%	15.4%

Table 8.3.1 Comparisons of the different identification methods for the glass tube process, T-S denotes the two-step method

These comparisons show that under the same experimental condition, the Markov parameter approach supplies the most compact model, while the two-step method gives the most accurate model of of the glass tube process.

Verifying the upper bound matrix

Another advantage of the two-step method is its ability to supply an upper bound matrix for the errors of the identified model. Before using the derived upper bound matrix for robust control, some tests must be performed to verify it. Because the true process model is never perfectly known, the test can only be done based on simulations. To do this, a model identified using the two-step method is used as the "process", the output error residuals of the model are used as the "disturbances". Two independent Gaussian white noises are used as the test signals; and the model is simulated using these inputs. The disturbances are added to the simulated outputs; the levels of the disturbances are adjusted so that the noise-to-signal ratios at the two outputs are both 10% (in power). 1000 samples are used in the test. We think that this represents the real experiment well.

Using the simulated input/output data, first a high order equation error model with structure [30, 30] is estimated, then a reduced model with the same structure as the "process" is obtained by model reduction. The errors of the high order model and the reduced model are plotted together with the upper bounds in Fig. 8.3.2. We see that the upper bounds cover the model errors in most of the frequencies, and that the reduced model has a higher accuracy. More tests of the upper bounds can be found in Zhu (1990).

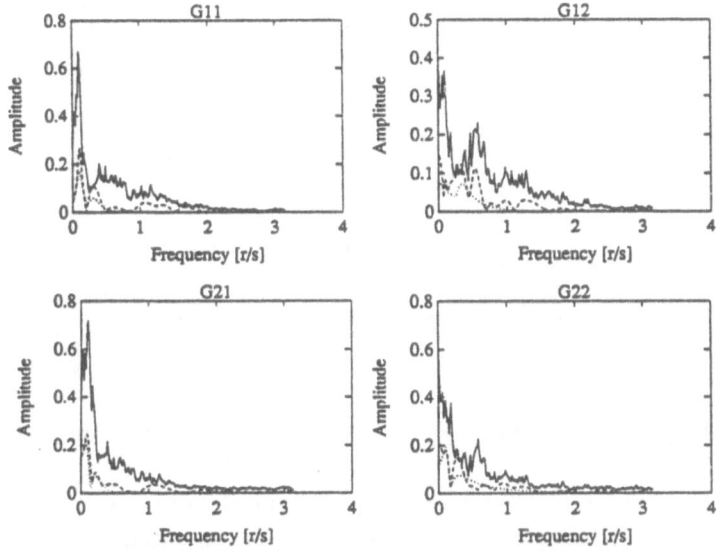

Fig. 8.3.2 Verifying the upper bounds. Error bounds (solid), errors of the high order model (dashed) and errors of the reduced model (dotted).

8.3.2 Identification of a Four-Effect Evaporator

Evaporators are used to reduce the water content of a product, for example, milk. Fig. 8.3.3 shows a four-effect evaporator which is used in milk powder production as a concentration process before the drying process. A multiple-effect evaporator is composed of similar effects. The product enters the effect and first reaches the distribution plate. This plate distributes the product to all evaporating tubes equally. The product flows as a film along the inside wall of the evaporation tubes while it is heated by condensing water vapour on the outside of the tubes. A two-phase flow of concentrate and vapour reaches the separator where product and vapour are separated. The product is pumped to the distribution plate of next effect and the vapour is directed to the next effect as well to condense on the outside of the evaporator tubes. The vapour from the last effect is condensed in a condenser and the product from this effect is directed to the drying stage.

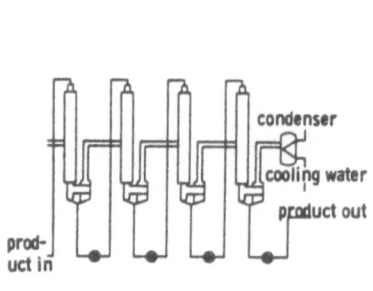

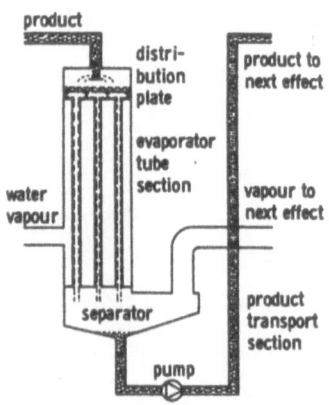

A) The complete process   B) One effect of the evaporator

Fig. 8.3.3 The four-effect evaporator (by courtesy of
Netherlands Institute for Diary Research (NIZO))

To obtain high-quality powder a constant dry matter content of the product is necessary. For energy saving in the drying process it is important to remove as much water as possible. In addition, reducing the start-up and shut-down times decreases the energy consumption and saves product. Both product quality and energy saving requirements need good control of the process. Due to the complex dynamic behaviour of the process it is very difficult for the operator to tune a PID controller. Hence identification is performed in order to obtain a model for advanced controller design or for automatically tuning a MIMO PID controller.

The outputs to be controlled are:

— output 1: dry matter content of the outcoming product;
— output 2: flow of outcoming product; and
— output 3: temperature of the outcoming product.

Three variables are chosen as control inputs:

— input 1: flow of feed;
— input 2: vapour flow to first evaporator stage;
— input 3: cooling water flow to condenser.

Other variables which also influence the outputs are dry matter content of feed, temperature of the feed, temperature of cooling water entering the condenser, and surrounding temperature. These are considered as disturbances. This arrangement results in a 3-input 3-output process with disturbances; see Fig 8.3.4.

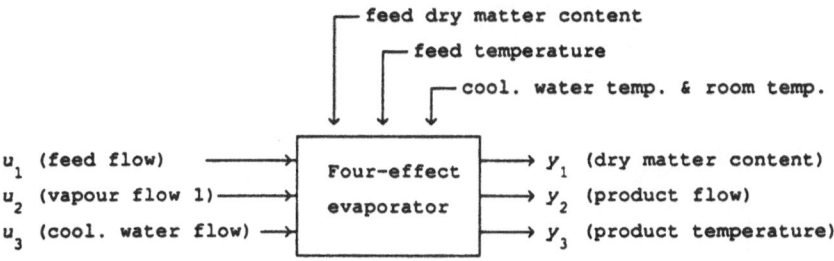

Fig. 8.3.4   Model the evaporator as a disturbed 3-input 3-output process

The staircase experiment has shown that the process is reasonably linear around given working points. Then PRBS test signals are used in the identification experiments. The input/output data for model estimation and model validation are collected from two different days, and each consists of 3000 samples. The two-step method is used to identify the process. The structure of the high order model is [30, 30, 30]; and reduced model has the structure [4, 5, 3]. The model is validated using the validation data, and the relative errors of the three outputs are 13.1%, 16.9%, and 12.2% respectively. Fig. 8.3.5 shows the result of model validation where only the first 1000 samples are plotted.

For comparison, the same data are used to estimate two diagonal ARMAX models with the structure [4, 5, 3] and [4, 5, 5] and a output error (OE) model with structrure [6, 6, 6]. The relative errors of model validations are given in Table 8.3.2. It is clear that the two-step method delivers a more compact model with higher accuracy.

Modelling by first principles, process identification and multivariable control for the evaporator can be found in van Wijck et. al. (1992) and Quaak et.al. (1992).

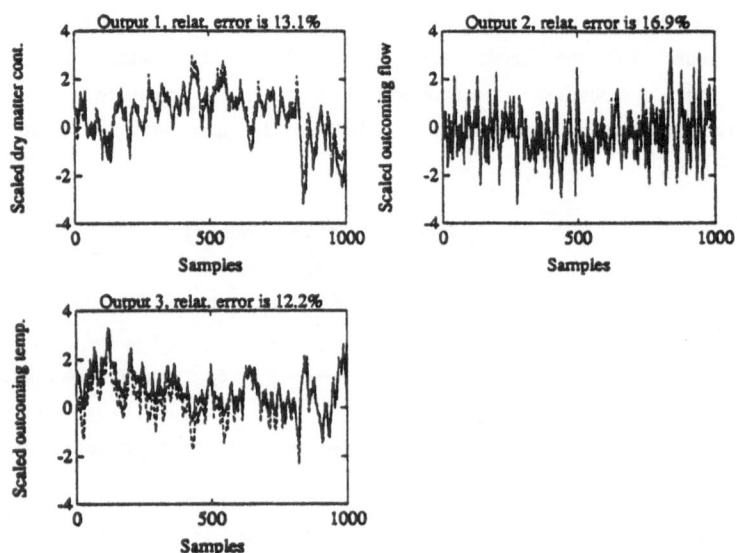

Fig. 8.3.5   Model validation for the four-effect evaporator.
"solid line" := measured output; "dashed line" := model output

T-S [4,5,3]	ARMAX [4,5,3]	ARMAX [4,5,5]	OE [6,6,6]
13.1%	16.7%	16.7%	28.0%
16.9%	17.4%	17.4%	21.8%
12.2%	37.8%	22.0%	19.1%

Table 8.3.2   Comparisons of the different identification methods for
the four-effect evaporator, T-S denotes the two-step method

For this process, both the ARMAX method and the two-step method need
about one hour computing time using PC-386 Matlab.

8.4   Closed Loop Identification of Coprime Factors

In practice it is sometimes advantageous or even necessary to do identification experiments in a closed loop under feedback, due to safety, economical
and technical reasons. If a process is unstable or nearly unstable, it is
obviously much safer to perform the identification tests in a closed loop.

Many industrial processes already have feedback PID controllers implemented; often they do not perform optimally (that is why we come to identify the process and design new controller), but they can safeguard the process during the identification experiments. The process manager and operator will not feel comfortable if their process is perturbed by identification test signals while the control loops are all open.

A closed loop system is shown in Fig. 8.4.1, where K is the feedback controller which stabilizes the process, G^o is the process, $u(t)$ is the process input vector, $y(t)$ is the output vector, $v(t)$ is the disturbance vector, $w_1(t)$ and $w_2(t)$ are the possible test signals, and $u_c(t)$ is the input vector of the controller.

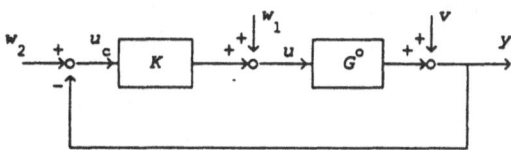

Fig. 8.4.1 Feedback control system

Normally, one tries to estimate a model of the open-loop process and design a controller based on the model. This can work well if the process is stable. If the process is unstable or nearly unstable, which is often the reason for closed loop experiment, difficulties arise in both the identification part and the (robust) controller design part. For the successful application of prediction error methods and our methods, the dynamics to be identified need to be stable. In order to apply robust control theory to the process model, e.g., to test the robust stability of a designed controller, it is necessary that the nominal model and the the process have the same number of unstable poles. This number is zero if the process and the identified model are both stable. If the process in open loop is unstable, it is difficult to guarantee that the model has the same number of unstable poles as the process. These problems can be avoided if the two-loop scheme proposed at the end of Chapter 3 is used. Another way to avoid these problems is to identify the models of the coprime factors of the unstable process and use them in control design. In fact, the models of the coprime factors are more suitable for robust controller design than the process model; see McFarlane and Glover (1989) and Schrama (1991).

The definitions of coprime factor descriptions are as follows (see

Vidyasagar, 1985).

— Denote N, D as stable transfer function matrices, then the pair (N, D) is called coprime if there exist stable transfer function matrices X, Y such that the Bezout identity is satisfied

$$XN + YD = I \qquad (8.4.1)$$

— (N, D) is a right coprime factorization of G if $\det D \neq 0$, $G = ND^{-1}$ and (N, D) is right coprime.

Analogously left coprime factorization can be defined. The coprimeness of N and D over stable systems means that these factors do not have common unstable zeros. This prevent hidden instabilities between N and D.

To see the idea of coprime factorization, we have a simple example:

— The (unstable) process

$$G(q) = \frac{1 - 0.7q^{-1}}{1 - 2q^{-1}} = \frac{q - 0.7}{q - 2}$$

— Stable factors

$$N = \frac{q - 0.7}{q - 0.5}, \qquad D = \frac{q - 2}{q - 0.5}$$

— Coprimeness. It is obvious that N and D are coprime because they do not have any common unstable zeros. To verify the Bezout identity, let $X = 15/13$ and $Y = -2/13$, then

$$XN + YD = (\frac{15}{13}) \frac{q - 0.7}{q - 0.5} + (\frac{-2}{13}) \frac{q - 2}{q - 0.5} = 1$$

Based on the properties of the stabilizing controller K, the following schemes are proposed for coprime factor identification.

A)   If K is stable

In this case, we use test signal $w_1(t)$ and let $w_2(t) = 0$. Then we have

$$\begin{cases} y(t) = N_1^o w_1(t) + v_1(t) \\ u(t) = D_1^o w_1(t) + v_2(t) \end{cases} \qquad (8.4.2)$$

where

$$\left\{\begin{array}{l} N_1^\circ = (I + G^\circ K)^{-1} G^\circ \\ D_1^\circ = (I + KG^\circ)^{-1} \end{array}\right. , \qquad \left\{\begin{array}{l} v_1(t) = (I + G^\circ K)^{-1} v(t) \\ v_2(t) = -(I + KG^\circ)^{-1} Kv(t) \end{array}\right.$$

We know that N_1° and D_1° are stable, because the closed loop system is stable; $v_1(t)$ and $v_2(t)$ are stationary stochastic processes because $v(t)$ is. It is not difficult to verify that $N_1^\circ (D_1^\circ)^{-1} = G^\circ$. Let $X = K$ (stable), $Y = I$, we have

$$\begin{aligned} XN_1^\circ + YD_1^\circ &= K(I + G^\circ K)^{-1} G^\circ + (I + KG^\circ)^{-1} \\ &= KG^\circ (I + KG^\circ)^{-1} + (I + KG^\circ)^{-1} \\ &= I \end{aligned} \qquad (8.4.3)$$

Therefore (N_1°, D_1°) *is a right coprime factorization of the process* G°. Note that in the second equality of (8.4.3) the matrix identity

$$(I + G^\circ K)^{-1} G^\circ = G^\circ (I + KG^\circ)^{-1}$$

has been used. In fact it can be shown that K being stable is also the necessary condition for the coprimeness of (N_1°, D_1°); see Zhu and Stoorvogel (1992). Hence we can state the following

Result 8.4.1: Suppose that K stabilizes G°. Then K is stable if and only if N_1° and D_1° are right coprime.

Now, based on the identification method of Section 8.2, we can estimate N_1° and the error bound matrix from the data pair $\{y(t), w_1(t)\}$; and estimate D_1° and the error bound matrix from the data pair $\{u(t), w_1(t)\}$. These are two open-loop identification problems because the test inputs $w_1(t)$ are not correlated with the disturbances $v_1(t)$ and $v_2(t)$. The Bezout identity (8.4.3) can be used for model validation of estimated coprime factors in the frequency domain.

B)   If $[K]^{-1}$ is stable (G° is square)

For industrial applications, most processes can be stabilized by stable controllers if designed properly. The widely used industrial PID controller is, however, on the verge of instability due to the integration. Hence other solutions are needed for processes with PID controllers. Assume that, for a given process, the number of inputs equals the number of outputs, and the PID controller is diagonal. Then the inverse of the PID controller is stable. In this case, we use test signal $w_2(t)$ and let $w_1(t) = 0$. Then we have

$$\begin{cases} y(t) = N_2^o w_2(t) + v_1(t) \\ u(t) = D_2^o w_2(t) + v_2(t) \end{cases} \tag{8.4.4}$$

where

$$\begin{cases} N_2^o = (I + G^o K)^{-1} G^o K \\ D_2^o = K(I + G^o K)^{-1} \end{cases}, \quad \begin{cases} v_1(t) = (I + G^o K)^{-1} v(t) \\ v_2(t) = -(I + KG^o)^{-1} Kv(t) \end{cases}$$

Similarly, we can see that (N_2^o, D_2^o) is a right coprime factorization of G^o with Bezout identity

$$N_2^o + K^{-1} D_2^o = I \tag{8.4.5}$$

Result 8.4.2: Suppose that K stabilizes G^o. Then K^{-1} is stable if and only if N_2^o and D_2^o are right coprime.

Therefore (8.4.4) defines two open loop identification problems for the right coprime factors of G^o; the Bezout identity (8.4.5) can be used for model validation.

C) **No restriction on the stabilizing controller K**

In this case, one can use the more general scheme of Schrama (1991) where it is necessary to factorize the controller K and a model that is stablized by the controller. Here, in order to retain the simplicity of cases A) and B), we suggest to identify a right coprime factorization of $G^o K$, and then design robust controller for $G^o K$ instead of for G^o. This can be done using test signal $w_2(t)$, let $w_1(t) = 0$ and *replace the process input $u(t)$ by the controller input $u_c(t)$*. Then we have

$$\begin{cases} y(t) = N_3^o w_2(t) + v_1(t) \\ u_c(t) = D_3^o w_2(t) - v_1(t) \end{cases} \tag{8.4.6}$$

where

$$N_3^o = (I + G^o K)^{-1} G^o K, \quad D_3^o = (I + G^o K)^{-1}, \quad v_1(t) = (I + G^o K)^{-1} v(t)$$

Then it is easy to see that (N_3^o, D_3^o) is a right coprime factor description of $G^o K$ with Bezout identity

$$N_3^o + D_3^o = I \qquad (8.4.7)$$

Hence we can identify N_3^o and D_3^o according to (8.4.6) and validate the model by the Bezout identity (8.4.7).

In this section we intend to bring two messages:

• The two-step method of Section 8.2 which was proposed for open-loop identification, can also be used to solve the MIMO closed loop identification problem. The trick is to identify the coprime factors of the process. We have shown that if the feedback controller is stable or its inverse is stable, this can be done easily.

• If the process is unstable, most of the existing identification methods cannot meet the needs of modelling for robust control; the problem can be avoided by identifying a pair of (stable) coprime factors of the unstable process.

Coprime factorization is central to the newly developed robust control theory; see Vidyasagar (1985). A reader who is not familiar with the theory may feel rather ambiguous about what we are talking about. There is no need to bother youself to understand all the material in this section except the two messages just mentioned. These (rather theoretical) part will become easier when you realy need them for solving practical problems.

8.5 Conclusions

In this chapter we have solved the problem of MIMO process identification as promised in the title of the book. The method is almost as simple as for SISO processes, which might be surprising to the reader. This is the consequence of the nice structure of the asymptotic theory, and open-loop assumption. The method has been tested on two industrial processes and the results have been compared with those of the prediction error methods. Besides the ability to supply error bounds in the frequency domain, it turns out that for these real industrial processes, under the same degree of model complexity, our method delivers more accurate models than the prediction error methods.

It is possible to develop a closed-loop version of the method; but it will be more complex. Instead, we have shown that the open-loop version of our method can be used for closed-loop experiments when the coprime factors of the process are identified.

In the next chapter we will present the results of identification and control of the glass tube manufacturing process.

CHAPTER 9

IDENTIFICATION AND ROBUST CONTROL OF THE GLASS TUBE PROCESS

One of the most important uses of identification is to achieve better control of a process. In fact many identification methods are developed by control engineers. Most of the work in the previous chapters was devoted to *identification for control*. Here we will shown the usefulness of identification techniques in process control by continuing the case study of the glass tube manufacturing process. First we outline a procedure from identification to control system design and test which we follow in practice (Section 9.1). Then the results of identification and control of the glass tube making process will be presented (Section 9.2). The purpose of this chapter is not to teach control theory. Instead, we try to convince our readers that identification is a very useful tool in the application of modern control theories.

9.1 From Identification to Robust Control; Guidelines

The following procedure is proposed, see Fig. 9.1.1, which reflects our view on the integration of identification and control. In the following, each step will be explained briefly.

1) Selection of the inputs and the outputs for identification

This is a crucial step in applications and most of the text books on identification and control do not discuss this topic thoroughly.

The choice of the input for identification was discussed in Chapter 3. In general two classes of inputs should be used for the identification, the control inputs and measurable disturbances. The use of measurable disturbances as inputs for identification will bring us the following benefits:

1) It can reduce the noise level in the identification, because the affects of the measurable disturbances on the outputs are identified and they no longer act as noises for identification.

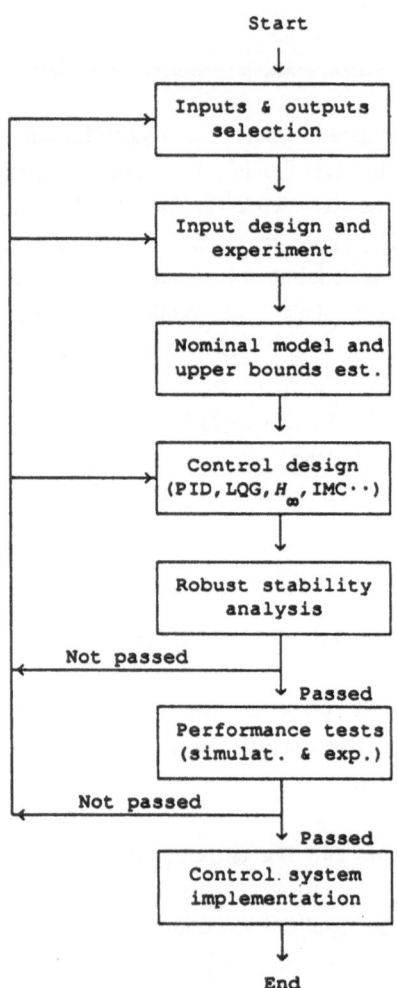

Fig. 9.1.1 From identification to robust control

2) The identified model from measurable disturbance to the outputs to be controlled can be used to design feedforward compensator. This will improve the control performance.

Suppose that all the variables to be controlled are easily measurable, choose them as the output and the is work done. In practice, however, the choice of the outputs for identification is not that trivial. This is mainly caused by the limitations of sensors and instrumentation. In a practical situation we can have the following types of outputs:

1) Outputs to be controlled and can be easily measured.

2) Outputs to be controlled, but so far cannot be measured.

3) Outputs to be controlled, but their measurements are difficult and it is too costly to use them for feedback control; examples are thermal couples for measuring very high temperatures (easily wearing out) and analyzers for measuring compositions (too slow).

4) Outputs not to be controlled but can be easily measured.

We should of cause use the first type of outputs as the outputs for identification; meanwhile we leave the second type of outputs to the measurement specialists. We may solve the measurement problem of the third type of outputs by using the fourth type of output as follows:

In the final identification experiment, measure all the first, third and fourth type of outputs. First identify two models from the measured data. The first model uses an extended set of inputs: the original identification inputs and the fourth type of outputs, its outputs are first and third type of outputs. The second model uses the fourth type of outputs as its inputs and the third type of outputs as outputs. Then there are two possibilities to solve the problem.

Method One. When the second model can simulate the third type of outputs as good as the first model, it means that the third type of outputs are totally determined by the fourth type of outputs. Thus we can replace the third type of outputs by the fourth type and control them. In this case, for the design of controller, a third model needs to be identified from the same final experiment whose inputs are the original identification inputs and outputs are the first and the fourth type of outputs.

Method Two. When the third type of the outputs cannot be fully determined by the fourth type, then design controller for both the first and third type of outputs as of they are all measurable during control. In the implementation stage, simulate the first model online and replace the sensor signals of the third type of outputs by those of the model.

The expensive measurements of the third type of outputs are only used to adjust the models. We see that with the help of MIMO process identification, we can go beyond the limit of sensors.

2)   Test input design and experiments

As explained in Chapter 3 and Chapter 7, this step consists of free-run experiment (for examining the influence of the disturbances), step experiments (for determining the dominant time constants), staircase experiments (for checking the static nonlinearity), white noise (or fast PRBS) experiments (for determining the bandwidths) and optimal input experiments (for model and error bound estimation). Each of these experiments will supply the necessary information about the process dynamics needed for the next step. Therefore, we recommend our readers to do them all if possible. It is, however, possible to skip some of the experiments, if the relevant information can be obtained from *a priori* knowledge and engineering experience.

A good way to realise the spectra of the optimal inputs is to filter white noise signals or fast PRBS signals.

3)   Nominal model and upper error bound estimation

There are many identification algorithms in the literature. If it is possible, try different methods and make good comparisons. Most of them cannot yet provide model error descriptions for use in robust controller design. In industrial applications, the most popular method is still the least-squares (LS) method. Besides the LS method, the output error method, the IV methods and the ARMAX model method are often used.

The methods developed in Chapters 6, 7 and 8 provide systematic solutions to industrial identification problems. Hence they are highly recommended for the user.

4)   Controller design

Many modern control techniques such as LQG (linear quadratic Gaussian) control, IMC (internal model control), *H*-infinity control can be applied at this step. Also PID control can be used here. With a good process model at hand, it is very easy to tune a PID controller for a MIMO process. Several controllers can be designed with different techniques; the best one will be determined by the following robust stability analysis and performance tests. This is not difficult if some relevant CAD packages are available.

An important remark we would like to make is about the use of feed-forward control for the compensation of measurable disturbances. The performance of feedback control is limited by the time delays that often exist in a industrial process, and by the constraints on the loop gains. So the effect of feedback regulation is to reduce the effect of slow distur-bances. A feedforward controller, on the other hand, can react on both slow and fast disturbances. The reason that feedforward control is not so widely used in conjuction with feedback control is the lack of a good process model that includes the measurable disturbances as its inputs. We believe that the application of process identification can change this situation.

5)   Robust stability test

The estimated upper bound matrix of the model errors obtained in Chapter 8 can be used for this test. The most suitable test criterion for our description of model uncertainty is the so called non-similarity scaling method which was developed by Kouvaritakis and Latchman (1985); see Zhu (1990).

6)   Performance tests by simulation and experiments

After having tested the stability robustness of a feedback control system the next logical step is to analyze the robust performance of the system. But the result of this kind of analysis will be very conservative. This is because, when the upper bound matrix is used, it is very hard to derive some sensible engineering measures of system performance without introducing con-siderable conservativeness for a MIMO system; cf. Lunze (1989). A more feasible approach is: (1) to check the nominal system performance by a simulation; and (2) to check the robust performance of the system by online tests of the control system. It is usually much cheaper and safer to perform this kind of experiment than the original identification experiments.

Remarks— This procedure is iterative. When a controller cannot pass the stability or the performance test, the expression of the upper bound matrix indicates directly or indirectly the ways for the redesign of the identification experiment and/or the re-estimation of the models and/or the redesign of the controller.

— In this procedure, the ultimate model validation is control performance tests, rather than the cross-validation at the identification step.

9.2 Identification and Control of the Glass Tube Manufacturing Process; Control Results

This process has been used several times for studying different identification methods. Now we will complete this case study by showing some results of model based control.

The identified model of the glass tube process is used to design a control system which has a internal model control structure; see Fig. 9.2.1. It is known that this scheme is equivalent to the Smith predictor control which can improve the tracking property for processes with large measurement delays. A good model is essential for the effectiveness of this scheme.

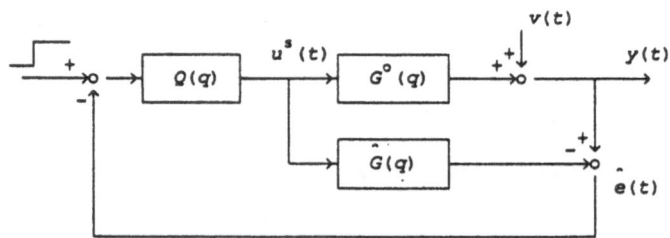

Fig. 9.2.1 Internal Model Control

The feedback controller Q is designed for good step response, decoupling and disturbance reduction. The process model is identified using the algorithm of Chapter 6. The upper bound matrix is calculated using the method of Chapter 8. In the identification stage, both white noise inputs and optimal inputs are used. The results show that the optimal input is superior to the white noise input for control performance. Upper bound matrices are estimated for the models. Fig. 9.2.2 shows the stability test of the system with the model obtained using the optimal input signals.

The achievement of the control system is remarkable: the change-over time has been reduced by a factor of more than 10 (from one hour to 5 minutes), which means the increase of flexibility; the standard deviations of the tube thickness and diameter have been reduced by 50%, which means an increase of product quality; see Fig. 9.2.3. Notice also that a setpoint change on the tube thickness does not disturb the diameter. In addition, these improvements imply considerable saving of energy.

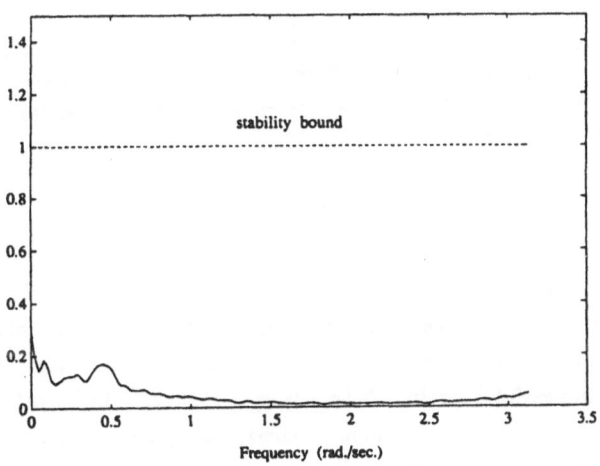

Fig. 9.2.2 Robust stability test of the control system

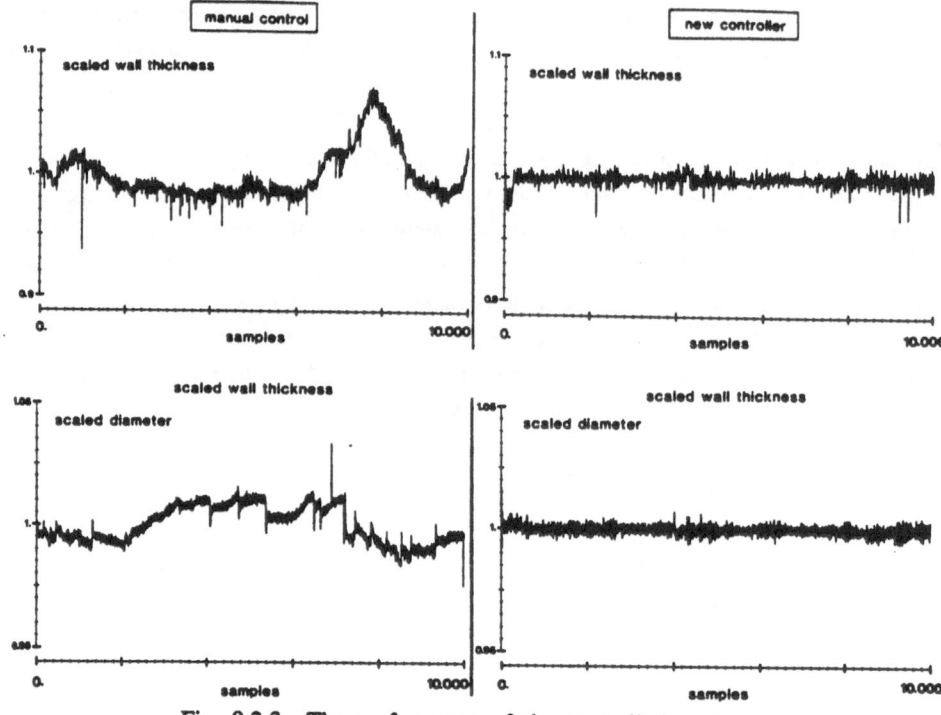

Fig. 9.2.3 The performance of the controlled system

9.3   Conclusions

In this chapter we have shown the use of process identification for process control. The realized computer control system of the glass tube process has brought significant improvement in production quality and flexibility. In fact, equally successful results have been obtained by our team in the identification and control of, among others, a glass feeder (glass industry) bulb making machines (lighting industry), a multi-effect evaporator (food industry) and a distillation column (refinery industry).

A compact and accurate process model is the key to apply modern control theory to industrial processes. In addition, the availability of an accurate process model can significantly improve the efficiency in tuning a classical PID controller, especially for a truly MIMO process.

CHAPTER 10

IDENTIFICATION FOR FAULT DIAGNOSIS; ESTIMATION OF CONTINUOUS-TIME MODELS

In the previous chapters we have studied the problem of identification for simulation and control. In this chapter we will study the problem of identification for fault diagnosis. Process fault diagnosis is part of *process supervision* which is a level higher than process control. Fault diagnosis consists of fault detection and fault isolation; the purpose of this is to improve the safety and reliability of the automated system. However, the implemented methods of fault diagnosis are still mainly based on simple limit value checking of some easily measurable signals or their derivatives.

Fault diagnosis based on modelling and estimation has received intensive attention in recent years, and is now entering industrial applications (Frank, 1990). These methods use mathematical models of the process to retrieve the information related to the faults. Depending on the faults to be detected there are basically two approaches of model based fault diagnosis: state estimation (Kalman filter or observer) method (Willsky, 1976) and parameter estimation method (Isermann, 1984). Both methods need accurate process models, hence identification and parameter estimation play an important role here.

For use in fault diagnosis continuous-time models should be estimated because their states or parameters are closely related to the physical parameters of the process. This motivates us to study the estimation of continuous-time models. Compared to the estimation of discrete-time models, the estimation of continuous-time models is an underdeveloped area. In order to exploit the lead of discrete-time identification, an indirect method of continuous-time identification is proposed and studied in Section 10.1. In this approach, first a discrete-time model is estimated by e.g. a prediction error method; then the parameters of the continuous-time model are determined according to the relation between the frequency response of the discrete-time model and that of the continuous-time model.

For a successful application of model based fault diagnosis, a precise process model is essential. Modelling errors (model uncertainty) can obscure the effect of faults and therefore are a source of false alarms. One way to solve the problem caused by modelling errors is to design detection methods that are robust or insensitive to modelling errors. Very recently, robust fault diagnosis has received more interest (Emami-Naeini et al., 1988 and Frank, 1990). A disadvantage of robust fault detection is that the algorithms are also insensitive to some of the faults, which implies missing alarms. To our opinion, robust fault diagnosis will be less effective than robust control, because feedback does not help here.

An alternative solution to the model uncertainty problem is to improve the model quality for the use in fault diagnosis. This direction, however, has not yet received much attention. Our idea is that in the process modelling and identification stage, the emphasis should be put on those model parameters that are related to the faults which are more likely to occur and difficult to detect. This requires an identification method which can enhance the accuracy of a subset of the total parameter vector to be estimated. To this end, an input design technique is proposed in Section 10.2. In Section 10.3 a simulation study is carried out and proposed methods are validated and compared with a direct method. Section 10.4 gives conclusions for this chapter. Only SISO processes are studied.

10.1 An Indirect Method of Continuous-time Model Estimation

There are basically two approaches to the identification of continuous-time models: direct approach and indirect approach; see Young (1981) and Unbehauen and Rao (1990) for surveys. In a direct method, derivatives (integrals) of the input/output signals are calculated by numerical derivative (integral) operations performed on the sampled input/output data, then the parameters of the continuous-time model are estimated from the calculated derivatives (integrals). The estimation method can be an LS method, an output error method, an IV method or an prediction error method. Compared to the estimation of discrete-time models, the direct method of continuous-time model estimation is more computationally demanding due to the necessity of numerical differential (integral) operations; these operations are also sensitive to disturbances and round-off noise; and the properties of the estimated models are less well understood.

The indirect approach is to first estimate a discrete-time model and then to transform the discrete-time model to a continuous-time model. The possible advantage of this indirect approach is that the estimation of discrete-time models is simple and its properties are well understood. The existing transformation methods require extensive computation, they are in part not straightforward, and their properties are not well known, so that they are not suitable for on-line real-time applications. Here an indirect method via discrete-time model estimation is proposed which is numerically simple and straightforward.

Denote the true transfer function as

$$G_c(s) = \frac{b_0 s^m + b_1 s^{m-1} + \cdots + b_m}{s^n + a_1 s^{n-1} + \cdots + a_n} \qquad m \le n \qquad (10.1.1)$$

where s is the variable of Laplace transform. The idea is to use the relation between the frequency response of the continuous-time model and that of the discrete-time model. Assuming that a zero-order hold is used when sampling the continuous-time process, which is often the case in applications, and also assuming the sampling frequency is much higher than the bandwidth of the process, we have (see, e.g., Isermann, 1989):

$$G_d(e^{i\omega T}) = \frac{1 - e^{-i\omega T}}{i\omega T} G_c(i\omega) \qquad \text{for } 0 \le \omega \le \frac{\pi}{T} \qquad (10.1.2)$$

where $G_d(e^{i\omega})$ and $G_c(i\omega)$ are the frequency responses of the discrete-time and continuous-time models respectively, T is the sample interval. This relation is precise if there is no aliasing. Based on this relation, the following procedure of indirect estimation is proposed:

Procedure 10.1.1

Step 1   Determining the order of the discrete-time model.

In fault diagnosis applications, an analytic model of the process is usually derived based on first principles. This model is given in continuous-time with some of the parameters to be estimated. Hence the structure (the orders m and n) can be determined according to the relation between s-transfer function and z-transfer function, assuming a zero-order hold at the input.

Step 2   Estimating the parameters of the discrete-time model.

The estimation of linear discrete-time models is well established. In this step, one can follow the prediction error method using a Box-Jenkins model structure. One can also use the two-step method studied in Chapter 7.

Step 3   Determining the parameters of the continuous-time model.

Denote $\hat{G}_d(i\omega)$ as the frequency response calculated from the discrete-time model. From (10.1.2), we obtain an estimate of the frequency response of the continuous-time model

$$\hat{G}_c(i\omega) = \frac{i\omega T}{1 - e^{-i\omega T}} \; \hat{G}_d(e^{i\omega T}) \quad \text{for } 0 \le \omega \le \frac{\pi}{T} \tag{10.1.3}$$

Now, (10.1.1) gives

$$\hat{G}_c(i\omega) = \frac{\hat{b}_0(i\omega)^m + \hat{b}_1(i\omega)^{m-1} + \cdots + \hat{b}_m}{(i\omega)^n + \hat{a}_1(i\omega)^{n-1} + \cdots + \hat{a}_n}$$

which is equivalent to

$$\hat{G}_c(i\omega)[(i\omega)^n + \hat{a}_1(i\omega)^{n-1} + \cdots + \hat{a}_n] = \hat{b}_0(i\omega)^m + \hat{b}_1(i\omega)^{m-1} + \cdots + \hat{b}_m \tag{10.1.4}$$

The parameters, $\hat{a}_1 \; \hat{a}_2 \; \cdots \; \hat{a}_n \; \hat{b}_0 \; \hat{b}_1 \; \cdots \; \hat{b}_m$, can be obtained by solving a set of linear equations generated by (10.1.4). Note that evaluating (10.1.4) at one frequency point will generate two real valued equations. Hence using $n+1$ frequencies will generate $2n + 2$ equations which are enough for determining $n + m + 1$ parameters.

The equation (10.1.4) should be evaluated at frequency points in low and middle frequency range of the process transfer function; because at high frequencies, some degree of aliasing is inevitable, which implies that relation (10.1.2) less precise at high frequencies.

The conversion from discrete-time to continuous-time is very simple and costs little computing time; hence this method is suitable for on-line estimation.

Remark— In the identification literature, the discrete-time to continuous-time conversion based on frequency response has been proposed long time ago; see, the survey of Unbehauen and Rao (1990). In these methods, however, the frequency response is estimated by some non-parametric methods; then the parameters of the continuous-time model are estimated from the raw data of the frequency response.

10.2 Enhancing a Subset of the Parameters by Input Design

For our purpose, one should chose input signals such that the generated input/output data sequence is most informative for the estimation of those parameters which are important for the purpose of fault diagnosis. Denote θ as the parameter vector of the process model. If the parameter estimates are consistent, the covariance matrix, $\text{cov}\hat{\theta}$, can be used to measure the accuracy of the estimates. Denote W as a diagonal, nonsingular matrix whose diadonal entries are positive real numbers. Then the input design can be expressed formally as

$$\min_{u \in U} [\text{tr}(W\text{cov}\hat{\theta})] \qquad (10.2.1)$$

where u is the input signal to be chosen, U is the set of allowed inputs. This is called A-optimality in the literature (Mehra, 1974). The variances of the parameters can be influenced by the weighting matrix W. This is, however, a very difficult numerical optimization problem. To avoid this difficulty, we will propose a frequency domain method.

   Assume that the process disturbance is white noise, the covariance matrix expression under prediction error method in the frequency domain is (Ljung, 1987)

$$\text{cov } \hat{\theta} \approx C_1 \left[\int_{-\pi}^{\pi} \frac{\partial G(e^{i\omega})}{\partial \theta} \frac{\partial G(e^{-i\omega})}{\partial \theta^T} \Phi_u(\omega) d\omega + C_2 \right]^{-1} \qquad (10.2.2)$$

where $G(e^{i\omega})$ is the true transfer function of the process, C_1 and C_2 are constants and $\Phi_u(\omega)$ is the spectrum of the input. This expression shows the mechanism of how the input spectrum influences the covariance matrix. To obtain a small covariance matrix, we should use the input energy at frequencies where the frequency response is sensitive to parameter variations. If a particular parameter or several parameters are of special interest, then vary them and check where the frequency response is most sensitive and put more input energy there. This hold for both discrete-time and continuous-time models (see Mehra, 1974).

   Based on the above discussion the following heuristic procedure is proposed for input design:

1) Derive a nominal model by mathematical modelling and parameter estimation using the method of Section 10.1, where the input can be a white noise. Denote the model as

$$\hat{G}(s) = \frac{\hat{B}(s)}{\hat{A}(s)} \qquad (10.2.3)$$

2) Select the parameters of interest. Generate the test input, $u(t)$, by filtering a white noise $e(t)$:

$$u(s) = \mu \, F(s)e(s) \qquad (10.2.4)$$

where s is the variable of Laplace transform; μ is a constant which is used to adjust the input such that the amplitude or the energy is limited. The filtering in (10.2.4) can be approximated by a digital filter.

The choice of the filter is the following: denote θ_i as the parameter of interest, then

$$F(s) = \left[\frac{\partial \hat{G}(s)}{\partial \theta_i} \right]^{\lambda} \qquad (10.2.5)$$

where λ is a real number to be determined. If two parameters, θ_i and θ_j, are of interest, then

$$F(s) = \alpha_i \left[\frac{\partial \hat{G}(s)}{\partial \theta_i} \right]^{\lambda} + \alpha_j \left[\frac{\partial \hat{G}(s)}{\partial \theta_j} \right]^{\lambda} \qquad (10.2.6)$$

where α_i and α_j are used to adjust the relative importance of the two parameters. The same idea holds for more parameters. The choice of λ is also important. According to some simulation studies, $\lambda = 1/2$ seems to be a good choice.

3) Perform a identification experiment using the input in (10.2.4) and estimate the parameters by the indirect method of Section 10.1.

10.3 A Simulation Study

The continuous-time true transfer function is taken from Sagara and Zhao (1990) who used a numerical integration approach to estimate continuous-time models (direct method). The transfer function is

$$G_c(S) = \frac{0.0s + 5.0}{s^2 + 2.8s + 4.0} \qquad (10.3.1)$$

In order to make a comparison, we generate the input/output data using the same conditions as used by Sagara and Zhao (1990), except the initial values which play no important role in discrete-time model estimation when the experiment time is sufficiently long. These conditions are:

— The input is a combination of three sinusoids

$$u(t) = \sin 0.714t + \sin 1.428t + \sin 2.142t$$

— The sampling interval $T = 0.05$s.

— The output disturbance $v(t)$ is Gaussian white noise. The noise-to-signal ratio is 20% which is defined as the STD (standard deviation) of the disturbance divided by the STD of the noise-free output (4% in power).

— The number of samples $N = 1000$ (50s experiment).

— The number of experiments is 20.

— The simulation is performed using a zero-order hold; the initial values in the simulation are zero.

Box-Jenkins method is used for the estimation of the discrete-time model. The results of our method are compared with the results of Sagara and Zhao (1990) who used an instrumental variable (IV) method for parameter estimation; see Table 10.3.1. We find that, for the given input signal, the result of the indirect method is very similar the result obtained by the direct method. Remember that the computing cost of our method is much lower than that of the direct method; because we do not need integration operations on the input/output data.

True model parameters	Result of Sagara & Zhao	Result of the indirect method
$a_1 = 2.8000$	2.768 ±0.094	2.7765 ±0.0973
$a_2 = 4.0000$	3.940 ±0.153	3.9480 ±0.1724
$b_1 = 0.0000$	0.037 ±0.071	-0.0171 ±0.0673
$b_2 = 5.0000$	4.922 ±0.199	4.9455 ±0.2059
	MSE=0.0890	MSE=0.0926

Table 10.3.1   Comparing the indirect and the direct methods

In the following we will show the effectiveness of the input design method proposed in Section 10.2. A white noise input is used to estimate a nominal model. A filtered white noise is used to enhance parameter a_2. In the simulation, the disturbance and the noise-to-signal ratio are the same as before. The same disturbance is added to the output when a filtered white noise input is used. The power of the filtered input is the same as the power of the white noise input. Also Box-Jenkins models are estimated. No prefilter is used here.

The results are shown in Table 10.3.2. We find that a input with a continuous spectrum (white noise or filtered white noise) is much better than the periodic signal for the identification; compare Table 10.3.1.

True model parameters	Nominal model	Enhancing parameter a_2
$a_1 = 2.8000$	2.8110 ±0.0369	2.8035 ±0.0208
$a_2 = 4.0000$	4.0046 ±0.0574	4.0016 ±0.0305
$b_1 = 0.0000$	0.0012 ±0.0192	-0.0008 ±0.0112
$b_2 = 5.0000$	5.0124 ±0.0747	5.0037 ±0.0378
	MSE=0.0109	MSE=0.0029

Table 10.3.2   Results of white noise input and filtered input

The input design makes a significant improvement in parameter estimation. We also see that the filtered input improves the accuracy of all the parameters. This is because, for this example, all the specific filters for enhancing various parameters are all similar lowpass filters.

10.4 Conclusions

In this chapter, the estimation of continuous-time models is studied for the purpose of fault diagnosis. First an indirect approach to continuous-time model estimation is proposed. The idea is simple and natural: transform the discrete-time model to a continuous-time model by fitting the frequency response. The calculation involved is easy and straightforward. It is, therefore, suitable for on-line estimation. Secondly, a heuristic frequency domain method of input design is proposed. In the simulation study, the indirect method is validated and compared with a recent direct method. Also the method of input design has been shown very effective. For readers who are interested in the applications of model based fault diagnosis, see the survey paper of Isermann (1984) where two examples are presented.

SYMBOLS AND ABBREVIATIONS

A General Mathematical Symbols

:=	is by definition equal to
$\approx$	is approximately equal to
$\otimes$	the Kronecker product
$\rightarrow$	tends to
$\lim_{n \to \infty}$	the limit as n tends to infinity
$\lim_{\delta \to 0}$	the limit as δ tends to zero
$\in$	is an element of
$\subset$	is the subset of
R^d	d dimensional real vectors
AsN	asymptotic normal distribution
E	expectation operator
$\arg \min_{\theta \in D} V(\theta)$	value of $\theta \in D$ that minimizes the loss function $V(\theta)$

B Matrices

A_{ij}	the (i,j) element of the matrix A
A^T	the transpose of the matrix A
A^{*}	the conjugate transpose of the matrix A
A^{-1}	the inverse of the matrix A
A^{-T}	the inverse of the transpose of the matrix A
$\lambda(A)$	the eigenvalue of the matrix A
$\sigma(A)$	the singular value of the matrix A
$det(A)$	the determinant of the matrix A
I, I_p	identity matrix, identity matrix with dimension $p \times p$
$colA$	column operator stacking the columns of the matrix A on top of each other

C Signals, processes and models

$u(t)$	m-dimensional input vector at time t
$y(t)$	p-dimensional output vector at time t
$v(t)$	p-dimensional output disturbance vector at time t

$\{e(t)\}$	p-dimensional white noise vector
z^N	input/output data sequence
$R_u(\tau)$	autocorrelation function matrix of $u(t)$
$\Phi_u(\omega)$	auto spectrum matrix of $u(t)$
P	process dynamics
$\{G^o_k\}_{k=1,\cdots,\infty}$	impulse response of the process
$\{G_k\}_{k=1,\cdots,\infty}$	impulse response of the process model
$\theta,\ \hat{\theta},\ \theta_o$	parameter vector, its estimate and the true value
q^{-1}	delay operator
$G_o(q)$	the transfer operator of the process
$G(q),\ \hat{G}(q)$	the model and the model estimate of the process transfer operator
$G_o(e^{i\omega})$	the transfer function (frequency response) matrix of the process
$G(e^{i\omega}),\ \hat{G}(e^{i\omega})$	the model and the model estimate of the process transfer function matrix
$C(q),\ C(e^{i\omega})$	the transfer operator and the transfer function matrix of the feedback controller
$\Delta(e^{i\omega})$	modelling errors of the transfer function estimates
$\bar{\Delta}(\omega)$	upper bound matrix of the modelling errors
$\|G(e^{i\omega})\|_\infty$	infinity norm of the transfer function matrix

D Abbreviations

w.p.	with probability
ARMA	Auto Regressive Moving Average
ARMAX	Auto Regressive Moving Average eXogeneous
ARX	Auto Regressive eXogeneous
FIR	Finite Impulse Response
IMC	Internal Model Control
MA	Moving Average
MIMO	Multi-Input Multi-Output
SISO	Single-Input Single-Output

REFERENCES

Åström, K.J. and T. Bohlin (1965). Numerical identification of linear dynamic systems from normal operating records. *IFAC Symposium on Self-Adaptive Systems*, Teddington, England.

Åström, K.J. and P. Eykhoff (1971). System identification - a survey. *Automatica*, Vol. 7, pp. 123-162.

Åström, K.J. and B. Wittenmark (1984). *Computer Controlled Systems: Theory and Design*. Prentice-Hall, Englewood Cliffs, N.J..

Backx, T. (1987). *Identification of an Industrial Process: A Markov Parameter Approach*. Dr. dissertation, Dept. EE., Eindhoven University of Technology, Eindhoven, The Netherlands.

Backx, A.C.P.M. and A.A.H. Damen (1989). Identification of industrial MIMO processes for fixed controllers—part 1 and part 2. *Journal A*, Vol. 30, pp. 3-12, and pp. 33-43.

Van den Boom, T., M. Klompstra and A. Damen (1991). System identification for H_∞-robust control. *Preprint of the 9th IFAC/IFORS Symposium on Identification and System Parameter Estimation*, Budapest, July, 1991. pp. 1431-1436.

Box, G.E.P. and G.M. Jenkins (1970). *Time Series Analysis, Forecasting and Control*. Holden-Day, San Francisco.

Clarke, D.W. (1967). Generalized least squares estimation of parameters of a dynamic model. *Proceedings of First IFAC Symposium on Identification in Automatic Control Systems*, Prague, paper 3.17.

Doyle, J.C. (1982). Analysis of feedback systems with structured uncertainty. *IEE Proceedings, Part D*, Vol. 129, pp. 242-250.

Doyle, J.C. and G. Stein (1981). Multivariable feedback design: concepts for a classical/modern synthesis. *IEEE Trans. Autom. Control*, Vol. AC-26, pp. 4-16.

Emami-Naeini, A., M.M. Akhter and S.M. Rock (1988). Effect of model uncertainty on failure detection: the threshold selector. *IEEE Trans. Autom. Control*, Vol. 33, pp 1106-1115.

Eykhoff, P. (1974). *System Identification: Parameter and State Estimation*. John Wiley & Sons, New York.

Frank, P.M. (1990). Fault diagnosis in dynamic systems using analytical and knowledge-based redundancy—a survey and some new results. *Automatica*, Vol. 26, pp. 459-474.

Gantmacher, F.R. (1957). *The Theory of Matrices. Vol. I.* Chelsea Publishing Company, New York.

Gerth, W. (1972). *Zur Minimalrealisierung von Mehrgrossenübertragungssystemen durch Markovparameter*, PhD dissertation, Fakultät für Maschinenwesen der Technischen Universität Hannover, Hannover, Germany.

Gevers, M. and L. Ljung (1986). Optimal experiment design with respect to the intended model application. *Automatica*, Vol. 22, pp. 543-554.

Glover, K. (1984). All optimal Hankel-norm approximations of linear multivariable systems and their L_∞-error bounds. *Int. J. Control*, Vol. 39, pp. 1115-1193.

Goodwin, G.C. and M.E. Salgado (1989). A stochastic embedding approach for quantifying uncertainty in the estimation of restricted complexity models. *Int. J. Adaptive Control and Signal Processing*, Vol.3, pp. 333-356.

Goodwin, G.C., M. Gevers and B. Ninness (1992). Quantifying the error in estimated transfer functions with application to model order selection. *IEEE Trans. Autom. Control.* Vol. 37, No. 7, pp. 913-928.

Grimble, M.J. (1986). Optimal H_∞ robustness and relationship to LQG design problems. *Int. J. Control*, Vol. 43, pp. 351-372.

Gunnarsson, S. (1988). *Frequency Domain Aspects of Modeling and Control in Adaptive Systems.* Dr. Dissertation, Dept. EE., Linköping University, Linköping, Sweden.

Gustavsson, I., L. Ljung and T. Söderström (1977). Identification of processes in closed loop–Identifiability and accuracy aspects. *Automatica*, Vol. 13, pp. 59-75.

Hansen, F.R., G.F. Franklin and R. Kosut (1989). Closed-loop identification via the fractional representation: experiment design. *Proc. of American Control Conf.*, pp. 1422-1427.

Helmicki, A.J., C.A. Jacobson and C.N. Nett (1989). H_∞ identification of stable LSI systems: a scheme with direct application to controller design. *Proc. Amer. of Contr. Conf.*, Pittsburg, pp. 1428-1434.

Hsia, T.C. (1977). *System Identification: Least-Squares mehtods.* Lexington Books, Lexington.

Isermann, R. (1984). Fault detection based modelling and estimation methods — a survey. *Automatica*, Vol. 20, pp. 387-404.

Isermann, R. (1989). *Digital Control Systems.* Springer-Verlag, Berlin.

Janssen, P.H.M. (1988). *On Model Parametrization and Model Structure Selection for Identification of MIMO Systems.* Dr. dissertation, Eindhoven University of Technology, Eindhoven, The Netherlands.

Jenkins, G.M. and Watts (1968). *Spectral Analysis and Its Applications.* Holden-Day, San Francisco.

Kailath T. (1980). *Linear Systems.* Prentice-Hall, Englewood Cliffs, N.J..

Kosut, R.L. (1986). Adaptive calibration: an approach to uncertainty modelling and on-line robust control design. *Proc. 25th Conf. Decision and Control*, Athens, Greece, December, 1986, pp. 455-461.

Kosut, R.L., M. Lau and S. Boyd (1990). Identification of systems with parametric and nonparametric uncertainty. *Proc. of American Control Conference*, San Diego, June 1990, pp. 2412-2417.

Kouvaritakis, B. and H. Latchman (1985). Necessary and sufficient stability criterion for systems with structured uncertainties: the major principal direction alignment principle. *Int. J. Control.* Vol. 42, pp. 575-598.

Lamaire, R.O., L.Valavani, M. Athans and G. Stein (1991). A frequency-domain estimator for use in adaptive control systems. *Automatica*, Vol. 27, No. 1, pp 23-38.

Lenssen, M. (1988). *Optimal Input Design for MIMO Processes.* Master Thesis, Dept. EE., Eindhoven University of Technology, Eindhoven, The Netherlands.

Ljung, L. (1985). Asymptotic variance expressions for identified black-box transfer function models. *IEEE Trans. Autom. Control,* Vol. AC-30, pp. 834-844.

Ljung, L. and Z.D. Yuan (1985). Asymptotic properties of black-box identification of transfer functions. *IEEE Trans. Autom. Control*, Vol. AC-30, pp.514-530.

Ljung, L. (1987). *System Identification: Theory for the User.* Prentice-Hall, Englewood Cliffs, N.J..

Lunze, J. (1984). Robustness tests for feedback control systems using multidimensional uncertainty bounds. *Syst. & Cont. Letters*, Vol. 4, pp. 85-89.

Lunze, J. (1989). *Robust Multivariable Feedback Control.* Prentice Hall, New York.

McFarlane, D. and K. Glover (1989). Robust controller design using normalized coprime factor plant descriptions. *In Lecture Notes in Control and Information Sciences*, Springer Verlag, 1989.

Mehra, R.K. (1974). Optimal input signals for parameter estimation in dynamic systems—survey and new results. *IEEE Trans. Autom. Control*, Vol. AC-19, pp. 753-768.

Moore, B.C. (1981). Principal component analysis in linear systems: controllability, observability and model reduction. *IEEE Trans. Autom. Control*, Vol. AC-26, pp. 17-32.

Morari, M. and E. Zafiriou (1989). *Robust Process Control*. Prentice-Hall, Englewood Cliffs, N.J..

Pernebo, L. and L.M. Silverman (1982). Model reduction via balanced state space representations. *IEEE Trans. Autom. Control*, Vol. AC-27, pp. 382-387.

Quaak, P., M.P.C.M. van Wijck and J.J. van Haren (1992). Comparison of process identification and physical modelling for falling-film evaporators. *IPCOS Report 92/190/MvW*, IPCOS b.v., Best, The Netherlands.

Sagara, S. and Z.-Y. Zhao (1990). Numerical integration approach to on-line identification of continuous-time systems. *Automatica*. Vol. 26, pp 62-74.

Schrama, R.J.P. (1991). An open-loop solution to the approximate closed-loop identification problem. *Preprints 9th IFAC/IFORS Symp. on Identification and System Parameter Estimation*, Budapest, 8-12 July, 1991, pp. 1602-1607.

Smith, O.J.M. (1958). *Feedback Control Systems*. McGraw-Hill Book Company, Inc., New York.

Söderström, T. and P. Stoica (1982). Some properties of the output error method. *Automatica*, Vol. 18, pp. 93-99.

Söderström, T. and P. Stoica (1989). *System Identification*. Prentice-Hall, New York.

Steiglitz, K. and L.E. McBride (1965). A technique for the identification of linear systems. *IEEE Trans. Autom. Control*, Vol. AC-10, pp. 461-464.

Sternad, M. (1987). *Optimal and Adaptive Feedforward Regulators*. Dr. Dissertation, Uppsala University, Uppsala, Sweden.

Stoica, P. and T. Söderström (1981). The Steiglitz-McBride algorithm revisited: convergence analysis and accuracy aspects. *IEEE Trans. Autom. Control*, Vol. AC-26, pp. 712-717.

Unbehauen, H. and G.P. Rao (1990). Continuous-time approaches to system identification—a survey. *Automatica*. Vol. 26, pp 23-35.

Vidyasagar, M. (1985). *Control System Synthesis: A Factorization Approach*. MIT Press, Cambridge, Massachusetts.

Wahlberg, B. (1989). Model reduction of high-order estimated models: the asymptotic ML approach. *Int. J. Control*, Vol. 49, No. 1, pp 169-192.

Wahlberg, B. and L. Ljung (1986). Design variables for bias distribution in transfer function estimation. *IEEE Trans. Autom. Control*, Vol. AC-31, pp. 134-144.

van Wijck, M.P.C.M, P. Quaak and J.J. van Haren (1992). Multivariable supervisory control of a four-effect falling-film evaporator. *IPCOS Report 92/181/MvW*, IPCOS b.v., The Netherlands.

Willsky, A.S. (1976). A survey of design methods for failure detection in dynamic systems. *Automatica*, Vol. 12, pp. 601-611.

Young, P. (1981). Parameter estimation for continuous-time models—a survey. *Automatica*. Vol. 17, pp 23-39.

Yuan, Z.D. and L. Ljung (1984). Black-box identification of multivariable transfer functions: asymptotic properties and optimal input design. *Int. J. Control*, Vol. 40, pp. 233-256.

Zadeh, L.A. (1962). From circuit theory to system theory. Proc. IRE, Vol. 50, pp. 856-865.

Zames, G. (1981). Feedback and optimal sensitivity: model reference transformations, multiplicative seminorms, and approximate inverses. *IEEE Trans. Autom. Control*, Vol. AC-26, pp. 301-320.

Zhu, Y.C. (1989a). Estimation of transfer functions: asymptotic theory and a bound of model uncertainty. *Int. J. Control*, Vol. 49, pp 2241-2258.

Zhu, Y.C. (1989b). Black-box identification of MIMO transfer functions: asymptotic properties of prediction error models. *Int. J. Adaptive Control and Signal Processing*, Vol. 3, pp. 357-373.

Zhu, Y.C. (1990) *Identification and Control of MIMO Industrial Processes: An Integration Approach*. Dr. Dissertation, Dept. EE, Eindhoven University of Technology, The Netherlands.

Zhu, Y.C., M.H. Driessen and P. Eykhoff (1990). Identification and control of servo systems. *Journal A*. Vol. 31, pp 34-41.

Zhu, Y.C. and A.C.P.M. Backx (1991). MIMO process identification for controller design: test signals, nominal model and error bounds. *Preprints of the 9th IFAC/IFORS Symposium on Identification and System Parameter Estimation*, Budapest, July, 1991, pp. 1202-1207.

Zhu Y.C. and A.C.P.M. Backx (1991). Increasing the performance of FDD via improved identification. *Preprints of IFAC/IMACS Symposium on Fault Detection, Supervision and Safety for Technical Processes*, Baden-Baden, September, 1991, pp 169-173.

Zhu Y.C., A.C.P.M. Backx and P. Eykhoff (1991). Multivariable process identification based on frequency domain measures. *Proceedings of 30-th IEEE Conference on Decision and Control*, Brighton, December, 1991, pp 303-308.

Zhu, Y.C. and A.A. Stoorvogel (1992). Closed loop identification of coprime factors. *Proceedings of 30-th IEEE Conference on Decision and Control*, Tucson, Arizona, December, 1992, pp 53-54.

INDEX

186